MACMILLAN/McGRAW-HILL

Math

Daily Practice Workbook with Summer Skills Refresher

Published by Macmillan/McGraw-Hill, of McGraw-Hill Education, a division of The McGraw-Hill Companies, Inc., Two Penn Plaza, New York, New York 10121.

Printed in the United States of America

1 2 3 4 5 6 7 8 9 079 08 07 06 05 04 03

Contents

Daily Practice

Summer Skills Refresher

Name ____________________

Addition Patterns • Algebra

Follow the rule. Complete each table.

1.

Rule: Add 3	
In	**Out**
2	5
4	____
6	____
8	____

2.

Rule: Add 2	
In	**Out**
3	____
5	____
7	____
9	____

3.

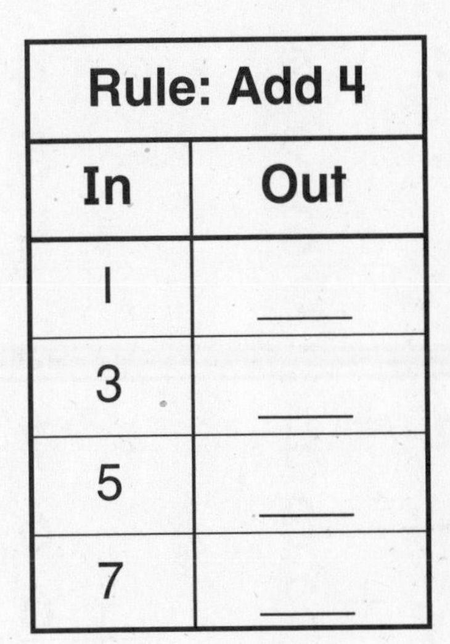

Rule: Add 4	
In	**Out**
1	____
3	____
5	____
7	____

4.

Rule: Add 1	
In	**Out**
5	____
6	____
7	____
8	____

5.

Rule: Add 0	
In	**Out**
2	____
3	____
4	____
5	____

6.

Rule: Add 5	
In	**Out**
0	____
2	____
4	____
6	____

7. Look at the table. What is the rule? ____________

In	Out
0	6
1	7
2	8
3	9

Name ______________________________

Add Three Numbers • Algebra

P 2-4 PRACTICE

Add.

1.	3	4	8	4	5	9
	2	5	0	3	4	1
	+ 3	+ 4	+ 2	+ 4	+ 6	+ 5
	8					

2.	4	7	9	8	7	5
	8	6	1	3	3	5
	+ 2	+ 6	+ 4	+ 8	+ 6	+ 5

3.	4	3	0	2	8	3
	6	5	7	4	2	6
	+ 8	+ 3	+ 7	+ 8	+ 3	+ 7

4.	6	4	8	5	1	3
	5	4	2	3	9	8
	+ 6	+ 7	+ 4	+ 5	+ 6	+ 2

Problem Solving

Solve.

5. Bao has 4 stamps. Tim has 9 stamps. Ben has 4 stamps. How many stamps are there in all?

______ stamps

6. There are 4 bear stamps, 6 wolf stamps, and 7 fox stamps. How many stamps are there in all?

______ stamps

Name ____

Problem Solving: Strategy

Draw a Picture

Draw a picture. Solve.

1. There are 4 kittens in the basket. 5 more kittens climb into the basket. How many kittens are in the basket?

   ___9___ kittens

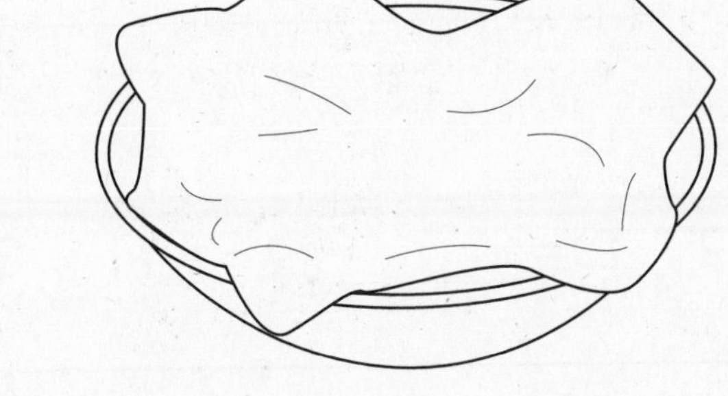

2. There are 7 boats on the lake. 6 more boats sail onto the lake. How many boats are on the lake?

   ______ boats

3. Joe finds 3 nests in the tree. Then he finds 4 more. How many nests did he find?

   ______ nests

Mixed Strategy Review

Solve.

4. Meg ate 5 grapes. Ray ate the same number. How many grapes did they eat in all?

   ______ grapes

5. **Write a problem** that you would draw a picture to solve. Share it with others.

   ____

   ____

Name ______________________________

Count Back to Subtract

P 3-1 PRACTICE

Count back to subtract.
You can use the number line.

0 1 2 3 4 5 6 7 8 9 10 11 12

1. $12 - 2 =$ ______ $11 - 3 =$ ______ $7 - 1 =$ ______

2. $8 - 3 =$ ______ $6 - 2 =$ ______ $10 - 2 =$ ______

3. $9 - 1 =$ ______ $7 - 3 =$ ______ $12 - 3 =$ ______

4. $10 - 1 =$ ______ $11 - 2 =$ ______ $8 - 2 =$ ______

5. ______ $= 12 - 1$ ______ $= 7 - 2$ ______ $= 10 - 3$

Problem Solving

6. There are 9 dogs in a pen.
3 dogs go home.
How many dogs are left?

______ dogs

7. There are 11 lions and 2 tigers at the zoo.
How many more lions than tigers are at the zoo?

______ lions

Name ____________________

Subtract All and Subtract Zero

P 3-2 PRACTICE

Subtract.

1. $7 - 1 = $ ___ $\quad 9 - 0 = $ ___ $\quad 8 - 8 = $ ___ $\quad 10 - 2 = $ ___

2. $11 - 11 = $ ___ $\quad 9 - 3 = $ ___ $\quad 6 - 0 = $ ___ $\quad 8 - 1 = $ ___

3. $12 - 1 = $ ___ $\quad 7 - 7 = $ ___ $\quad 9 - 2 = $ ___ $\quad 10 - 1 = $ ___

4. $8 - 3 = $ ___ $\quad 9 - 9 = $ ___ $\quad 7 - 2 = $ ___ $\quad 8 - 0 = $ ___

Problem Solving

Solve.

5. 10 children play ball.
All 10 go back to class.
How many children keep playing ball?

______ children

6. 8 girls take a walk.
No one leaves.
How many girls are left taking a walk?

______ girls

Name ____________________

Relate Addition to Subtraction

P 3-3 PRACTICE

Complete each number sentence.

1. 8 + 5 = 13 | 6 + 8 = ____ | 6 + 7 = ____

13 − 5 = ____ | 14 − 8 = ____ | 13 − 7 = ____

2. 4 + 9 = ____ | 8 + 8 = ____ | 6 + 9 = ____

13 − 4 = ____ | 16 − 8 = ____ | 15 − 6 = ____

3. $\begin{array}{r} 3 \\ +\,8 \\ \hline \end{array}$ $\begin{array}{r} 11 \\ -\,8 \\ \hline \end{array}$ $\begin{array}{r} 4 \\ +\,8 \\ \hline \end{array}$ $\begin{array}{r} 12 \\ -\,8 \\ \hline \end{array}$ $\begin{array}{r} 7 \\ +\,7 \\ \hline \end{array}$ $\begin{array}{r} 14 \\ -\,7 \\ \hline \end{array}$

4. $\begin{array}{r} 8 \\ +\,7 \\ \hline \end{array}$ $\begin{array}{r} 15 \\ -\,8 \\ \hline \end{array}$ $\begin{array}{r} 9 \\ +\,7 \\ \hline \end{array}$ $\begin{array}{r} 16 \\ -\,9 \\ \hline \end{array}$ $\begin{array}{r} 8 \\ +\,9 \\ \hline \end{array}$ $\begin{array}{r} 17 \\ -\,8 \\ \hline \end{array}$

5. $\begin{array}{r} 5 \\ +\,9 \\ \hline \end{array}$ $\begin{array}{r} 14 \\ -\,5 \\ \hline \end{array}$ $\begin{array}{r} 3 \\ +\,9 \\ \hline \end{array}$ $\begin{array}{r} 12 \\ -\,3 \\ \hline \end{array}$ $\begin{array}{r} 9 \\ +\,9 \\ \hline \end{array}$ $\begin{array}{r} 18 \\ -\,9 \\ \hline \end{array}$

Problem Solving

Solve.

6. There are 16 stamps. Pete uses 8 of the stamps. How many stamps are left?

____ stamps

7. Megan wrote 4 letters on Monday. She wrote 9 letters on Tuesday. How many letters did Megan write?

____ letters

Use with Grade 2, Chapter 3, Lesson 3, pages 37–38.

Name ______________________________

Missing Number • Algebra

Find each missing number.

1. $3 + \boxed{9} = 12$ $\quad 14 - 7 = \square$ $\quad \square + 8 = 14$

2. $15 - \square = 8$ $\quad 6 + \square = 11$ $\quad 13 - 9 = \square$

3. $4 + 7 = \square$ $\quad 12 - \square = 4$ $\quad 6 + 9 = \square$ $\quad 14 - 8 = \square$ $\quad 7 + 7 = \square$ $\quad 15 - 7 = \square$

4. $\square + 5 = 13$ $\quad 16 - \square = 8$ $\quad 6 + 7 = \square$ $\quad 17 - \square = 8$ $\quad \square + 9 = 14$ $\quad 14 - \square = 9$

5. $7 + \square = 10$ $\quad 11 - \square = 5$ $\quad \square + 8 = 13$ $\quad 18 - 9 = \square$ $\quad \square + 4 = 11$ $\quad 15 - \square = 8$

6. $\square + 4 = 12$ $\quad 10 - \square = 4$ $\quad 6 + 6 = \square$ $\quad 16 - \square = 7$ $\quad 8 + \square = 17$ $\quad 16 - 9 = \square$

Problem Solving

Solve.

7. Jeff has 10 stamps. He gets 10 more. How many stamps does he have now?

_______ stamps

8. Gina has 15 postcards. 7 are from the United States. How many are not from the United States?

_______ postcards

Name ____________________

Names for Numbers • Algebra

Write some ways to make the number at the top.

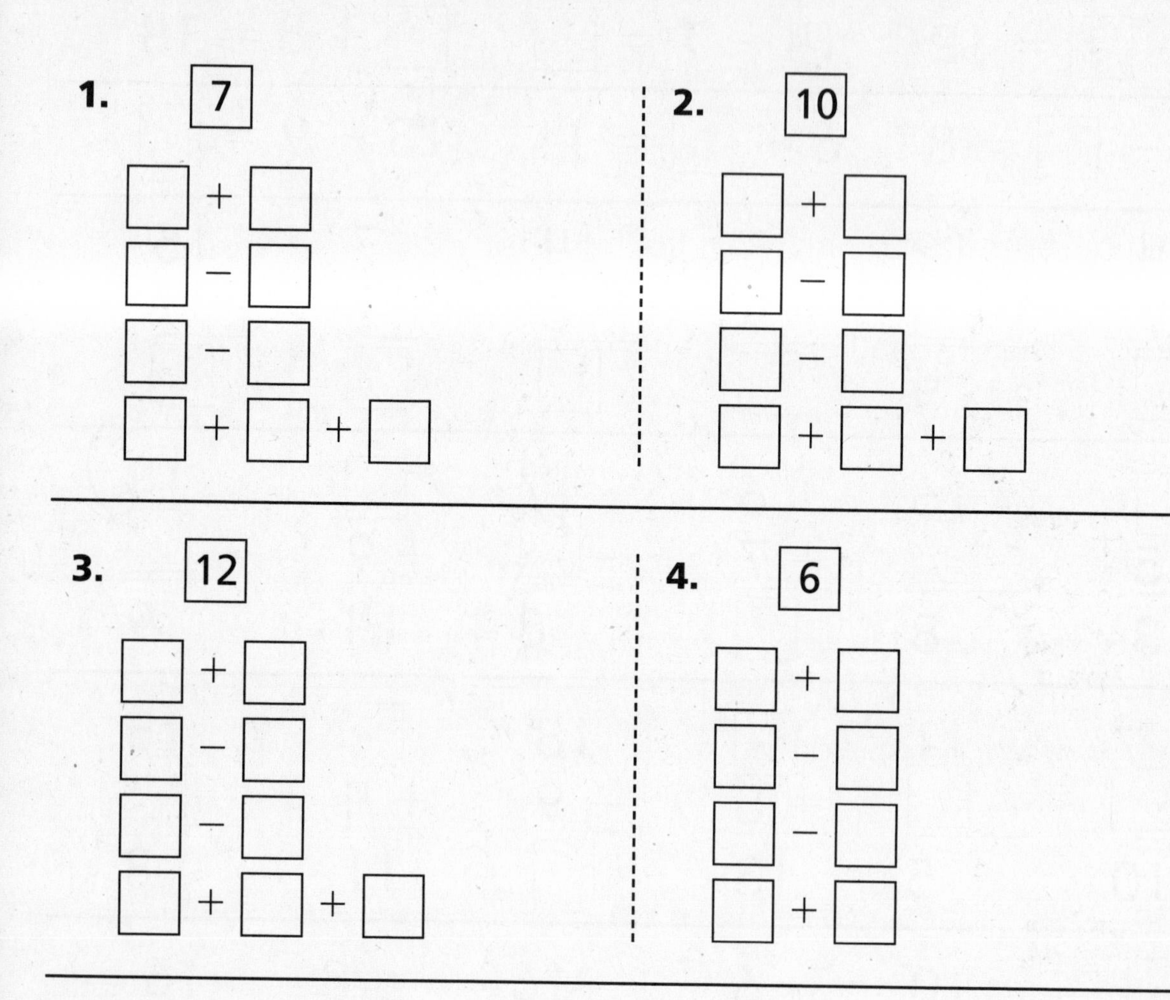

Circle your answer.

5. Which is not equal to 5?

4 + 1

8 − 3

5 + 5

7 − 2

6. Which is the same as 3 + 6?

15 − 6

8 + 4

16 − 8

1 + 7

Name ______________________________

Problem Solving Skill: Reading for Math

P 3-6 PRACTICE

Compare and Contrast

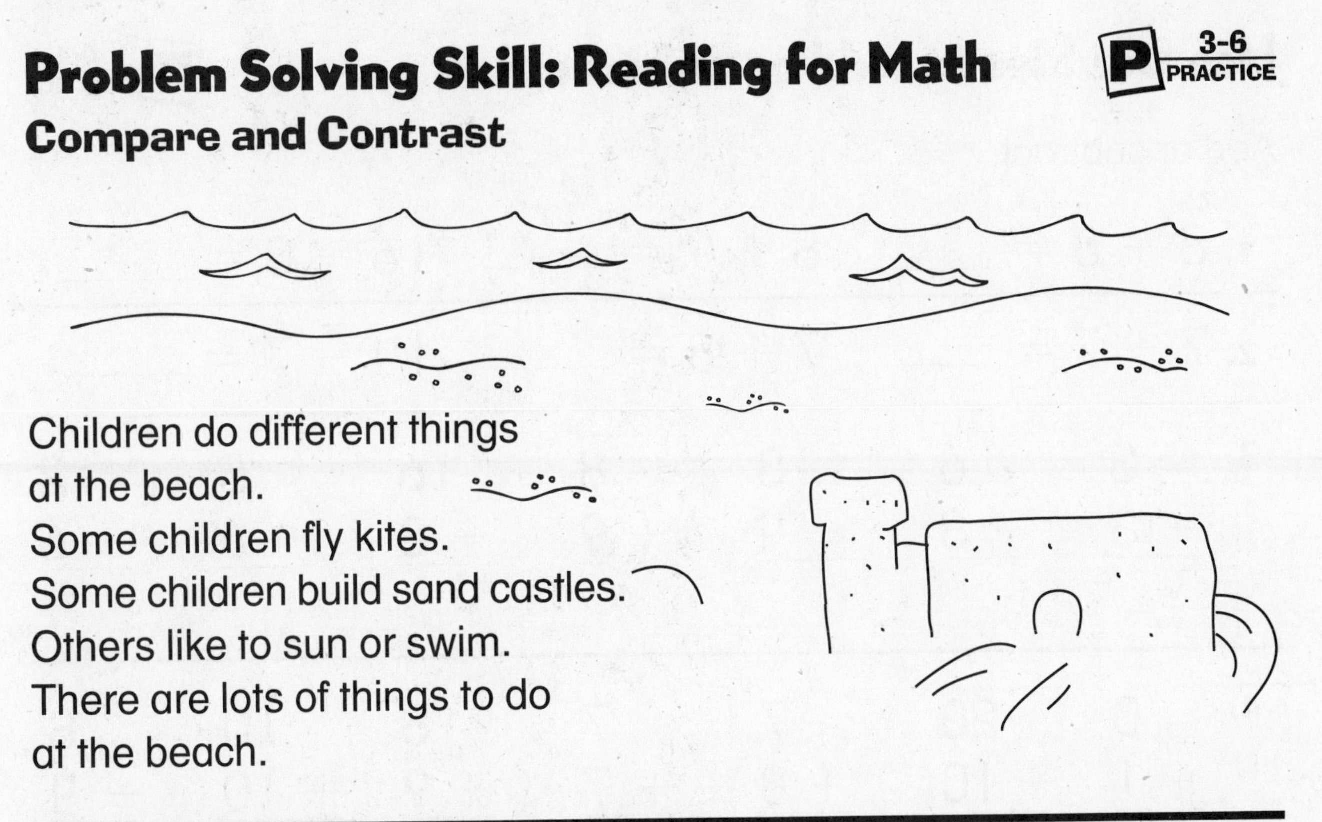

Children do different things
at the beach.
Some children fly kites.
Some children build sand castles.
Others like to sun or swim.
There are lots of things to do
at the beach.

Choose two things to do at the beach.
Tell how they are alike and different.

____

____

____

1. 8 children are making sand castles. 3 children leave to walk along the shore. Write a subtraction sentence to compare the two groups.

_____ − _____ = _____

_____ children are still making sand castles.

2. Frankie took 12 pictures of his friends at the beach. Mary took 7 pictures. How many more pictures did Frankie take than Mary took?

_____ pictures

Name ______________________________

Use Doubles to Add or Subtract

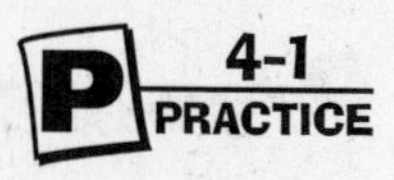

Add or subtract.

1. $8 + 8 =$ 16 $\quad 8 + 9 =$ 17 $\quad 16 - 8 =$ 8

2. $7 + 7 =$ _____ $\quad 7 + 8 =$ _____ $\quad 14 - 7 =$ _____

3.

3	6	8	8	16	5	12
+ 3	− 3	− 4	+ 9	− 8	+ 6	− 6

4.

0	20	7	7	8	10	8
+ 1	− 10	+ 8	+ 7	+ 9	+ 10	− 4

5.

5	18	4	14	2	10	4
+ 5	− 9	+ 5	− 7	+ 3	− 5	− 2

6.

0	1	6	13	8	17	5
− 0	+ 2	+ 7	− 6	+ 7	− 8	+ 5

Problem Solving

Solve.

7. A flower has 18 petals. 9 of the petals fall off. How many petals are left?

_____ petals

Show Your Work

Name ______________________________

Use 10 to Add and Subtract 9

Cross out 9 to subtract.

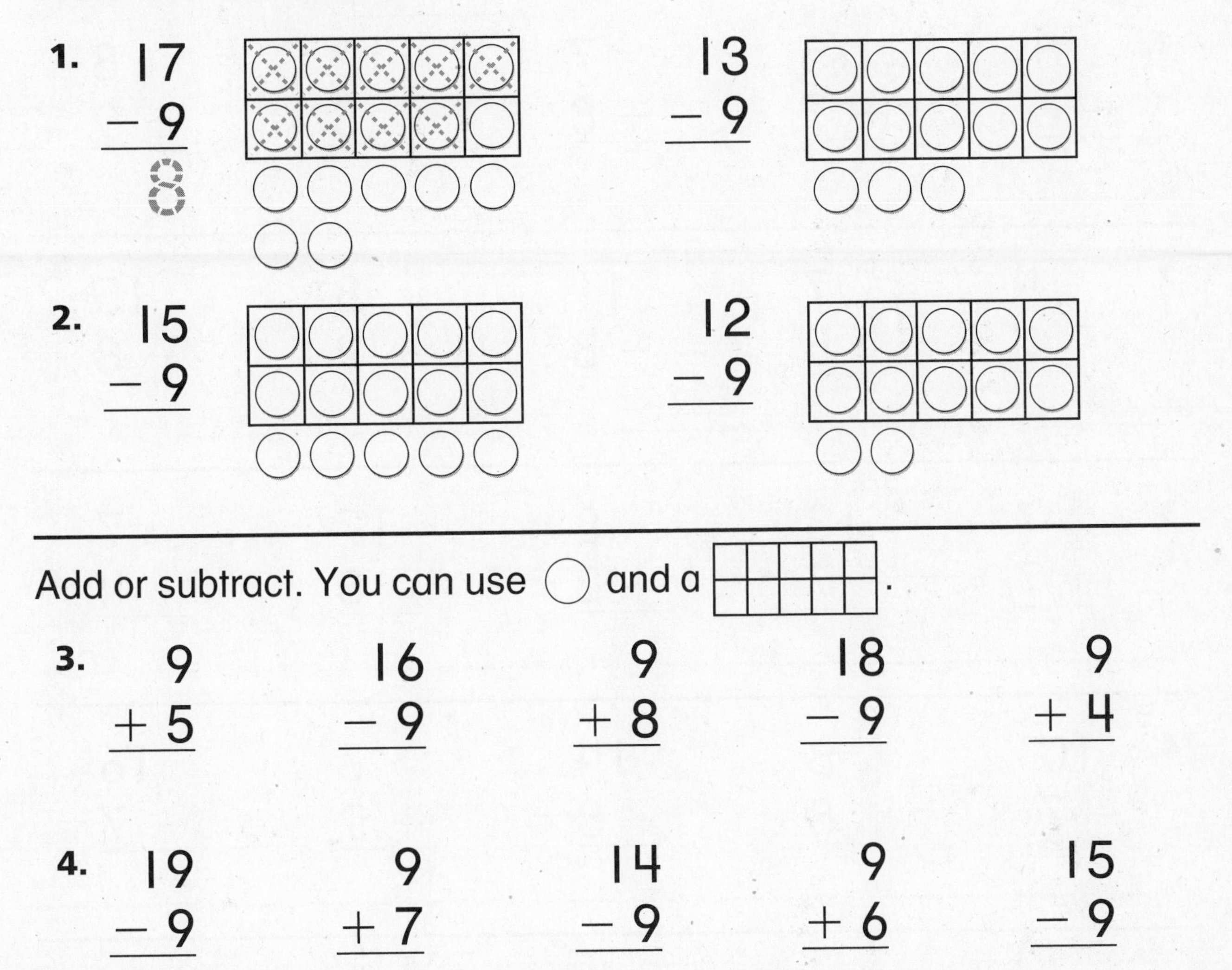

1. $\begin{array}{r} 17 \\ -\ 9 \\ \hline 8 \end{array}$ $\qquad$ $\begin{array}{r} 13 \\ -\ 9 \\ \hline \end{array}$

2. $\begin{array}{r} 15 \\ -\ 9 \\ \hline \end{array}$ $\qquad$ $\begin{array}{r} 12 \\ -\ 9 \\ \hline \end{array}$

Add or subtract. You can use ◯ and a ten-frame.

3. $\begin{array}{r} 9 \\ +\ 5 \\ \hline \end{array}$ $\quad$ $\begin{array}{r} 16 \\ -\ 9 \\ \hline \end{array}$ $\quad$ $\begin{array}{r} 9 \\ +\ 8 \\ \hline \end{array}$ $\quad$ $\begin{array}{r} 18 \\ -\ 9 \\ \hline \end{array}$ $\quad$ $\begin{array}{r} 9 \\ +\ 4 \\ \hline \end{array}$

4. $\begin{array}{r} 19 \\ -\ 9 \\ \hline \end{array}$ $\quad$ $\begin{array}{r} 9 \\ +\ 7 \\ \hline \end{array}$ $\quad$ $\begin{array}{r} 14 \\ -\ 9 \\ \hline \end{array}$ $\quad$ $\begin{array}{r} 9 \\ +\ 6 \\ \hline \end{array}$ $\quad$ $\begin{array}{r} 15 \\ -\ 9 \\ \hline \end{array}$

Problem Solving

Solve.

5. Huang caught 9 fish.
Lee caught 3 fish.
How many fish did they catch in all?

_____ fish

Show Your Work

Name ____________________

Use 10 to Add and Subtract 7 and 8

4-3 PRACTICE

Add or subtract. You can use ◯ and a ten-frame.

1.

8	15	7	12	8
+ 6	− 7	+ 8	− 8	+ 9
14				

2.

14	7	11	8	15
− 7	+ 5	− 8	+ 2	− 8

3.

7	13	8	17	7
+ 7	− 7	+ 3	− 8	+ 4

4.

10	8	14	7	16
− 7	+ 8	− 8	+ 9	− 7

Problem Solving

Solve.

Show Your Work

5. 12 people are on a bus.
8 people get off the bus.
How many people are still on the bus?

_____ people

Use with Grade 2, Chapter 4, Lesson 3, pages 57–58.

Name ______________________________

Fact Families

Complete each fact family.

1. Triangle: 14, 8, 6

$8 + 6 =$ 14
$6 + 8 =$ ____
$14 - 8 =$ ____
$14 - 6 =$ ____

2. Triangle: 13, 9, 4

$9 + 4 =$ ____
$4 + 9 =$ ____
$13 - 9 =$ ____
$13 - 4 =$ ____

3. Triangle: 17, 9, 8

$8 + 9 =$ ____
$9 + 8 =$ ____
$17 - 8 =$ ____
$17 - 9 =$ ____

4. Triangle: 13, 8, 5

$5 + 8 =$ ____
$8 + 5 =$ ____
$13 - 5 =$ ____
$13 - 8 =$ ____

5. Triangle: 15, 8, 7

$8 +$ ____ $= 15$
____ $+ 8 = 15$
$15 -$ ____ $= 7$
$15 - 7 =$ ____

6. Triangle: 16, 9, 7

____ $+ 7 = 16$
$7 +$ ____ $= 16$
$16 - 9 =$ ____
$16 -$ ____ $= 9$

7. Triangle: 14, 7, 7

____ $+ 7 = 14$
$14 - 7 =$ ____

8. Triangle: 18, 9, 9

____ $+ 9 = 18$
$18 -$ ____ $= 9$

Problem Solving

Solve.

9. Alex writes a fact family with the numbers 4 and 7. What number could complete the fact family?

____

10. David writes a fact family with the numbers 9 and 14. What number could complete the fact family?

____

Name ______________________________

Problem Solving: Strategy

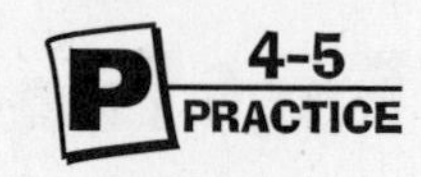

Write a Number Sentence

Solve. Write a number sentence.

Draw or write to explain.

1. Kelly buys 7 apples and 8 oranges. How many pieces of fruit does she buy altogether?

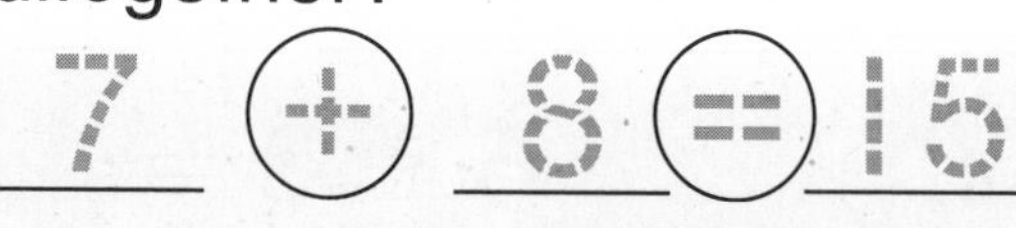

pieces of fruit

2. Max caught 11 fish. He threw 3 fish back. How many fish did Max keep?

____ ◯ ____ ◯ ____

fish

3. Rachel has 8 stickers. Jenny has 6 more stickers than Rachel has. How many stickers does Jenny have?

____ ◯ ____ ◯ ____

stickers

4. Sue caught 8 fish before lunch. She caught 4 more fish after lunch. How many fish did Sue catch altogether?

____ ◯ ____ ◯ ____

fish

Name ____________________

Tens

Write how many.

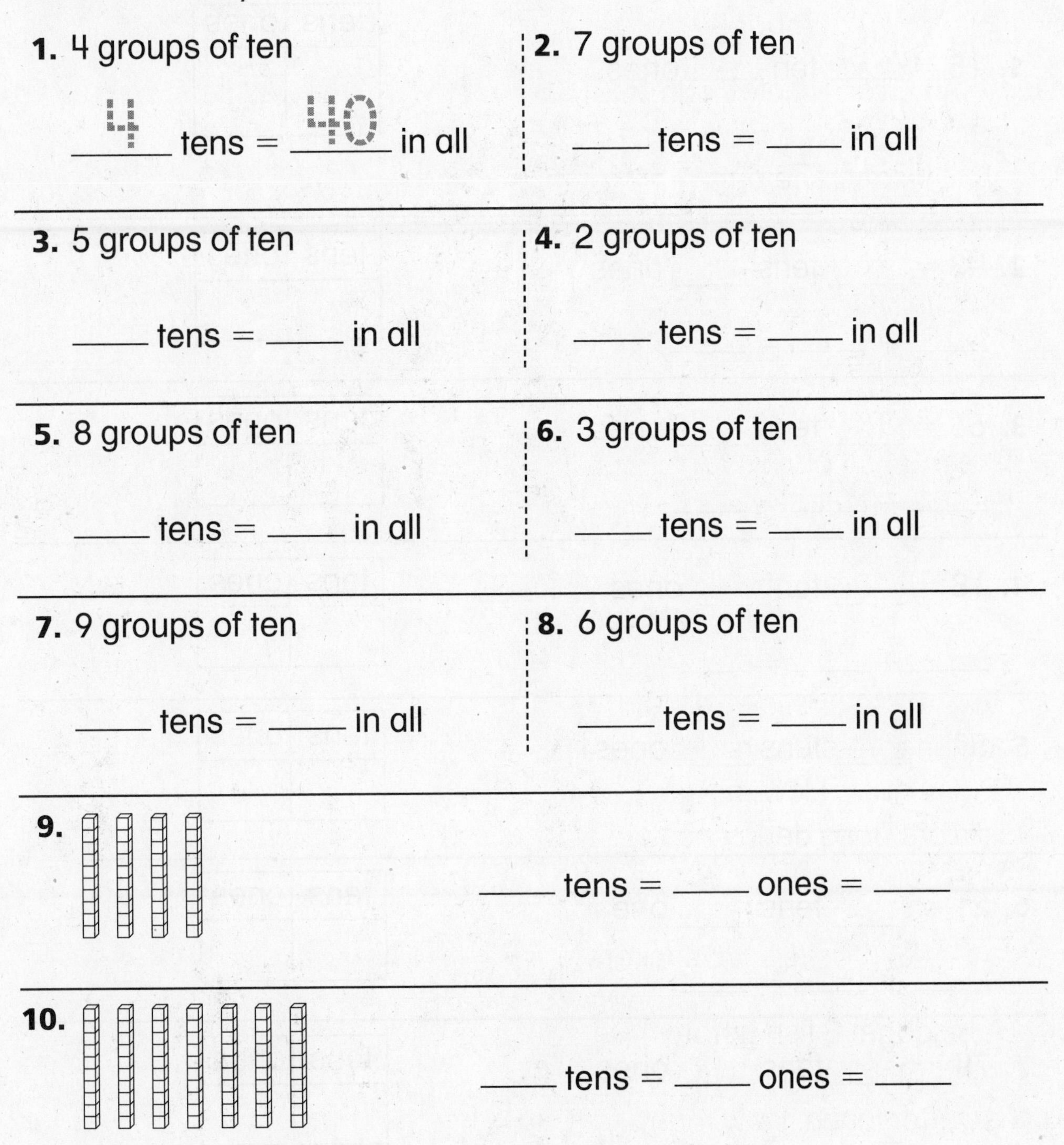

1. 4 groups of ten

4 tens = 40 in all

2. 7 groups of ten

____ tens = ____ in all

3. 5 groups of ten

____ tens = ____ in all

4. 2 groups of ten

____ tens = ____ in all

5. 8 groups of ten

____ tens = ____ in all

6. 3 groups of ten

____ tens = ____ in all

7. 9 groups of ten

____ tens = ____ in all

8. 6 groups of ten

____ tens = ____ in all

9.

____ tens = ____ ones = ____

10.

____ tens = ____ ones = ____

Name ______________________________

Tens and Ones

5-2 PRACTICE

Write how many tens and ones.

1. 15 = 1 ten 5 ones

10 + 5 = 15

tens	ones
1	5

2. 43 = ____ tens ____ ones

____ + ____ = ____

tens	ones

3. 66 = ____ tens ____ ones

____ + ____ = ____

tens	ones

4. 18 = ____ ten ____ ones

____ + ____ = ____

tens	ones

5. 59 = ____ tens ____ ones

____ + ____ = ____

tens	ones

6. 21 = ____ tens ____ one

____ + ____ = ____

tens	ones

7. 74 = ____ tens ____ ones

____ + ____ = ____

tens	ones

8. 32 = ____ tens ____ ones

____ + ____ = ____

tens	ones

Name ____________________

Place Value to 100

5-3 PRACTICE

Circle the value of the **bold** digit.

1. 63 — 6 or (60)
2. 48 — 8 or 80
3. 19 — 1 or 10
4. 86 — 8 or 80
5. 27 — 7 or 70
6. 71 — 7 or 70
7. 59 — 9 or 90
8. 15 — 5 or 50
9. 93 — 9 or 90
10. 41 — 1 or 10
11. 52 — 5 or 50
12. 76 — 6 or 60
13. 31 — 3 or 30
14. 29 — 2 or 20
15. 65 — 5 or 50

Write the number.

16. ______

17. ______

Name ____________________

Read and Write Numbers

Write each word as a number.

1. seventy 70
2. sixteen ____
3. thirty-seven ____
4. twenty-five ____
5. eighty-nine ____
6. twelve ____
7. forty-eight ____
8. ninety-two ____
9. fifty-one ____
10. sixty-three ____

Write each number word.

11. 23 ____________
12. 45 ____________
13. 78 ____________
14. 53 ____________
15. 13 ____________
16. 90 ____________
17. 84 ____________
18. 29 ____________
19. 35 ____________
20. 18 ____________
21. 59 ____________
22. 86 ____________
23. 31 ____________
24. 66 ____________

Name ____________________

Estimate Numbers

About how many notes are in each group?
Circle your estimate.

1.

about 20   about 50

2.

about 30   about 80

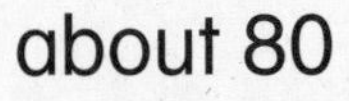

3.

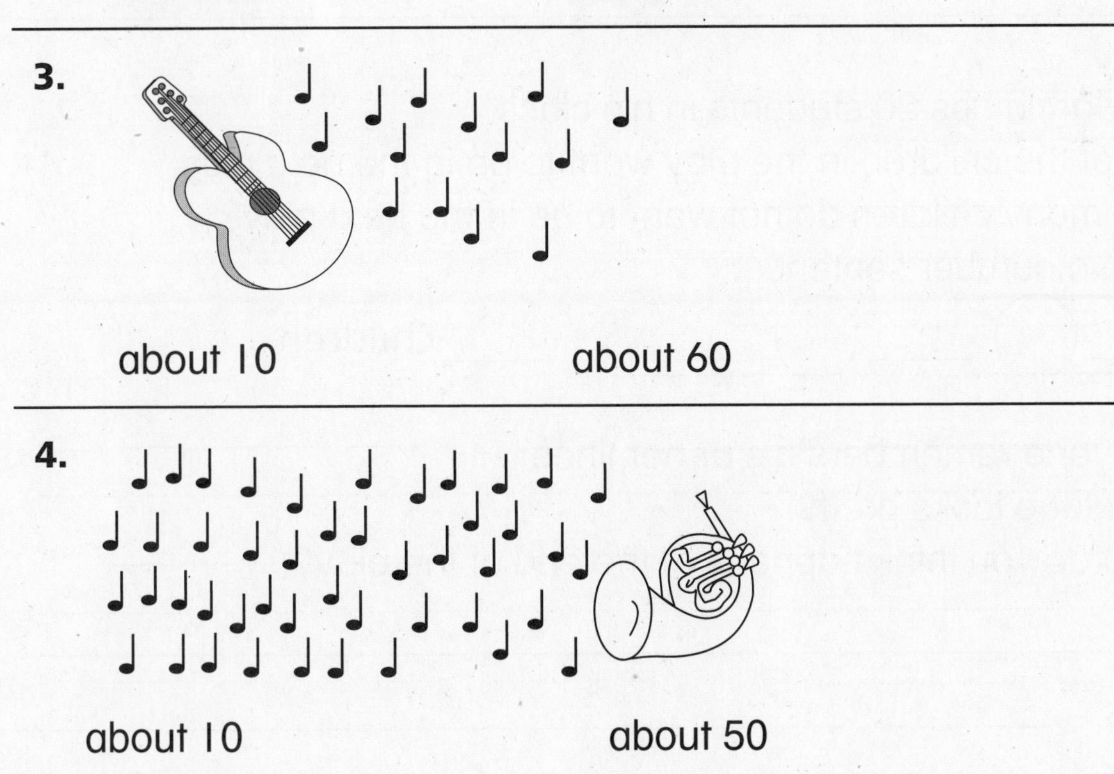

about 10   about 60

4.

about 10   about 50

Name ____

Problem Solving Skill: Reading for Math

Make Predictions

1. 50 students come to see a play.
Then 8 more students arrive.
Now how many students are at the play?

58 students
____

2. Play practice lasts 45 minutes.
The actors practice the entire play.
How long do you think the play is?

____

3. Mr. Garcia says, "Now 80 people are here!"
His class made 95 programs.
Are there enough programs for everyone?

____

4. Mr. Garcia has 20 students in his class.
Ten of the children in the play want to be in the next play.
How many children do not want to be in the next play?
Write a number sentence.

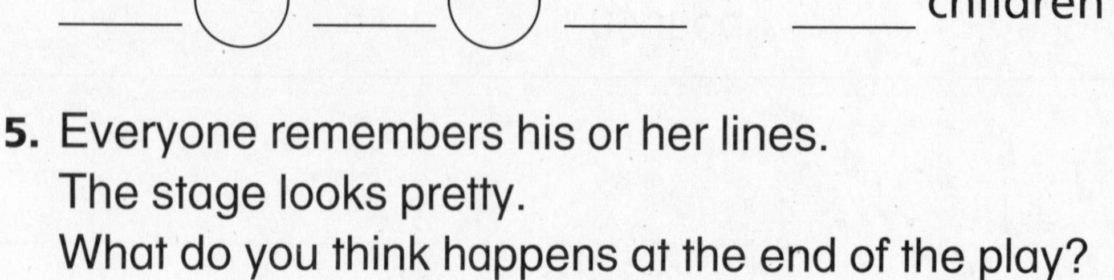

____ ◯ ____ ◯ ____ ____ children

5. Everyone remembers his or her lines.
The stage looks pretty.
What do you think happens at the end of the play?

____

____

Use with Grade 2, Chapter 5, Lesson 6, pages 87–88.

Name ____

Compare Numbers

Compare. Write > , <, or =.

1.	47 > 38	51 ◯ 45	19 ◯ 29
2.	36 ◯ 36	63 ◯ 72	23 ◯ 29
3.	95 ◯ 59	43 ◯ 49	78 ◯ 83
4.	31 ◯ 38	66 ◯ 6	45 ◯ 45
5.	27 ◯ 47	58 ◯ 81	49 ◯ 37
6.	83 ◯ 43	76 ◯ 57	58 ◯ 95
7.	28 ◯ 21	76 ◯ 69	40 ◯ 40
8.	80 ◯ 59	47 ◯ 59	68 ◯ 89
9.	33 ◯ 77	50 ◯ 60	32 ◯ 37
10.	16 ◯ 61	64 ◯ 82	95 ◯ 67
11.	28 ◯ 18	55 ◯ 34	47 ◯ 47
12.	92 ◯ 92	78 ◯ 87	36 ◯ 16

Name ____________________

Order Numbers

Write the number that comes just before.

1. 44, 45	____, 40	____, 78
2. ____, 61	____, 19	____, 54
3. ____, 24	____, 86	____, 20
4. ____, 73	____, 40	____, 48

Write the number that comes just after.

5. 50, ____	17, ____	99, ____
6. 42, ____	87, ____	13, ____
7. 29, ____	80, ____	39, ____
8. 27, ____	59, ____	66, ____

Write the number that comes between.

9. 29, ____, 31	91, ____, 93	55, ____, 57
10. 60, ____, 62	77, ____, 79	84, ____, 86
11. 48, ____, 50	39, ____, 41	15, ____, 17

Name ____

Skip-Counting Patterns

P 6-3 PRACTICE

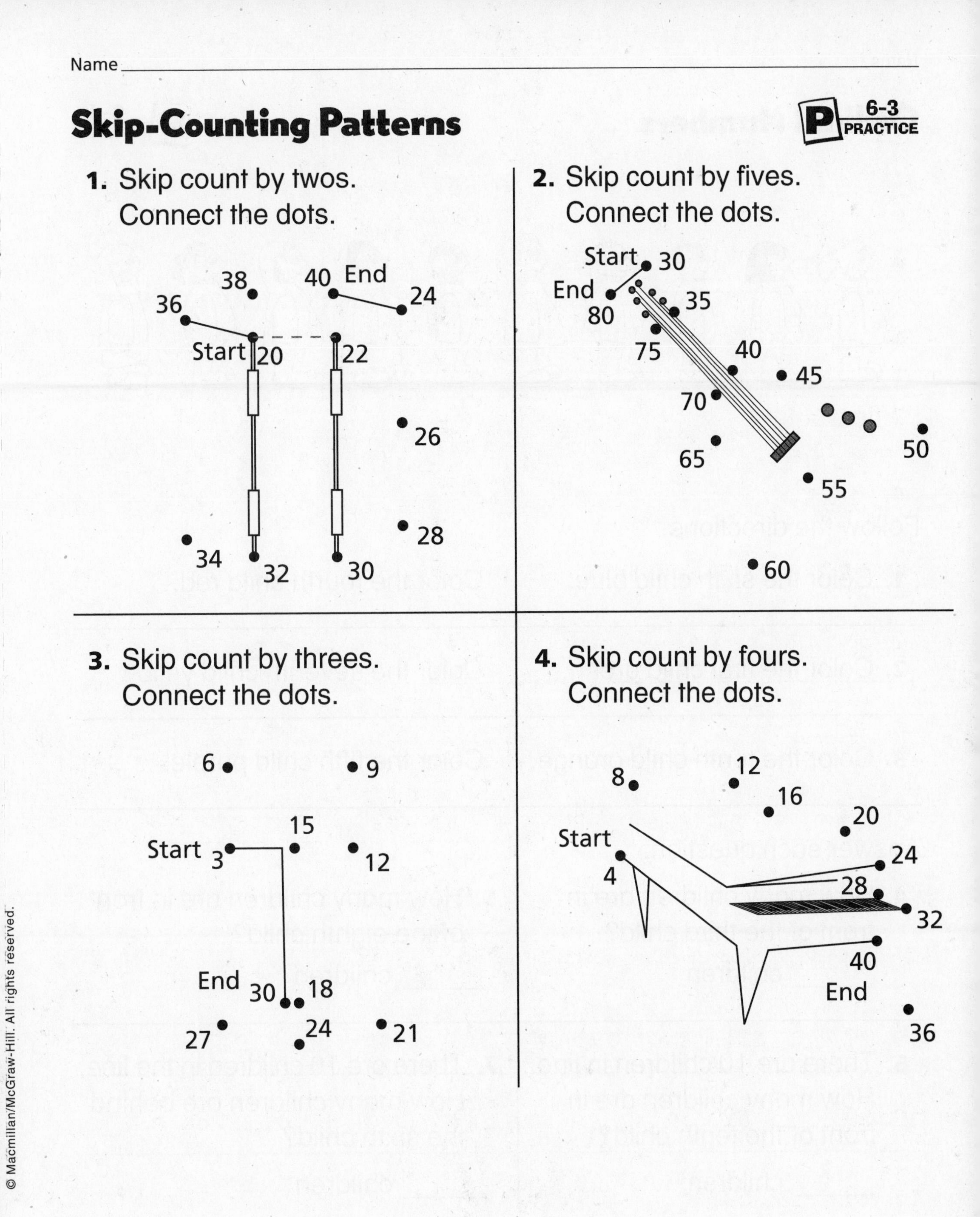

Name ______________________________

Ordinal Numbers

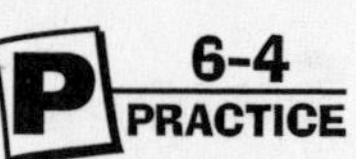

first

Follow the directions.

1. Color the sixth child blue. | Color the fourth child red.

2. Color the first child green. | Color the seventh child yellow.

3. Color the tenth child orange. | Color the fifth child purple.

Answer each question.

4. How many children are in front of the third child?

   _____ children

5. How many children are in front of the eighth child?

   _____ children

6. There are 10 children in line. How many children are in front of the tenth child?

   _____ children

7. There are 10 children in the line. How many children are behind the sixth child?

   _____ children

Name ______________________________

Even and Odd Numbers

Write if the number is even or odd.

1. 13 odd	45 odd	68 even
2. 33 ______	70 ______	28 ______
3. 44 ______	95 ______	67 ______
4. 72 ______	9 ______	79 ______

Write the next three even numbers.

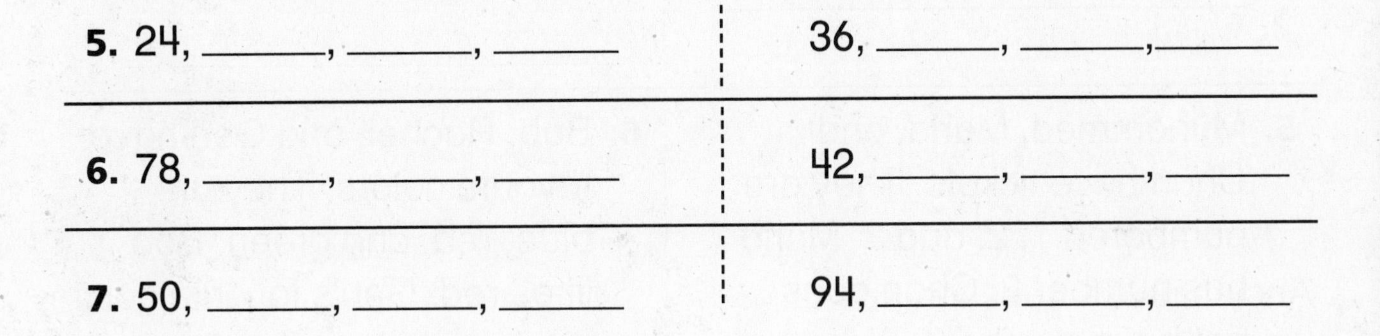

5. 24, ______, ______, ______	36, ______, ______, ______
6. 78, ______, ______, ______	42, ______, ______, ______
7. 50, ______, ______, ______	94, ______, ______, ______

Write the next three odd numbers.

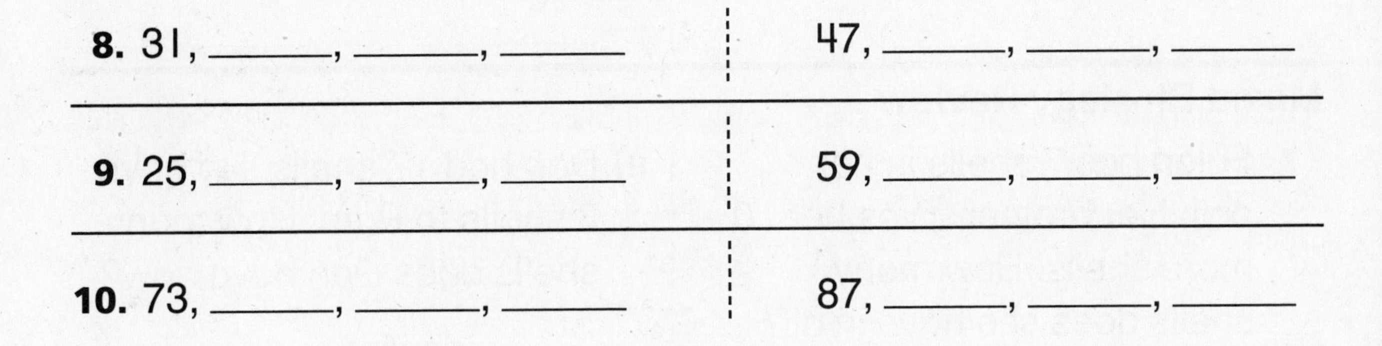

8. 31, ______, ______, ______	47, ______, ______, ______
9. 25, ______, ______, ______	59, ______, ______, ______
10. 73, ______, ______, ______	87, ______, ______, ______

Name ____

Problem Solving: Strategy

Use Logical Reasoning

Use logical reasoning to solve.

1. Zach, Alex, and Jen are on stage. Zach is on the left. Jen is not next to Zach. Who is in the middle?

____

2. Bob, Darryl, and Joe play kickball. Bob is up first. Joe does not go third. Who goes third?

____

3. Lori, Sara, Jill, and Ann are in line. Lori is first. Sara is after Lori. Ann is before Jill. Who is fourth?

____

4. Billy, Joanie, Neil, and Donna are waiting to see the nurse. A boy will go first. Donna goes second. Neil goes fourth. When does Joanie go?

____

5. Muhammed, Maria, and Chan have tickets. They are numbered 1, 2, and 3. Maria has number 2. Chan does not have number 3. Who has number 3?

____

6. Bob, Rachel, and Geri have favorite colors. They are blue, red, and green. Bob likes red. Geri's favorite color starts with the same letter as her name. What is Rachel's favorite color?

____

Mixed Strategy Review

7. Falon has 7 shells in her pail. Her brother gives her 10 more shells. How many shells does she have in all?

______ shells

8. Dan had 17 shells. He gave 3 shells to Pam. How many shells does Dan have now?

______ shells

Name ___

Pennies, Nickels, and Dimes

Use coins.
Show each price in two different ways.
Draw the coins.

Name ____________________

Count Coin Collections

Count to find the total amount.
Do you have enough money to buy the charm?
Circle *yes* or *no*.

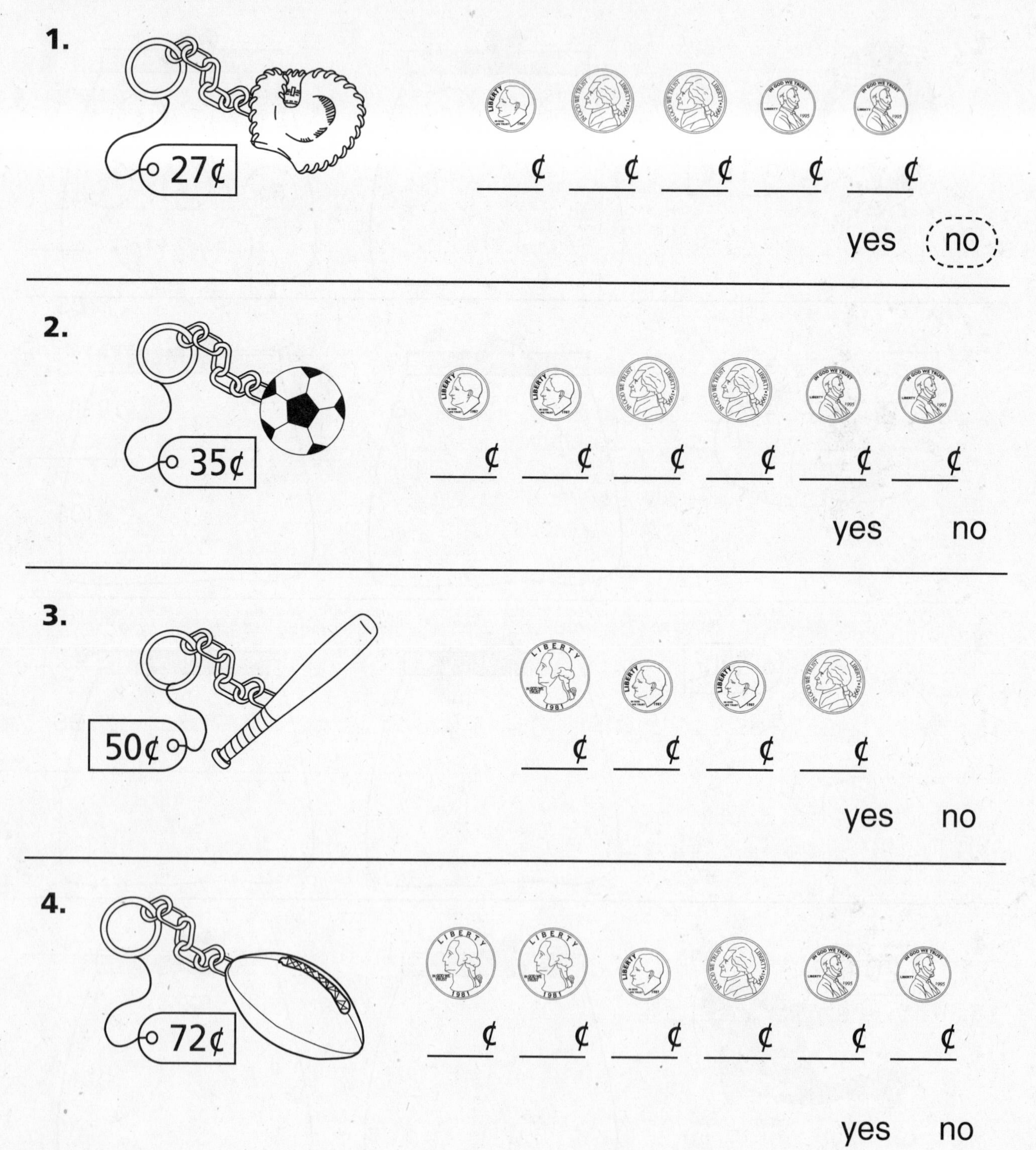

Name ____________________

Money and Place Value

7-3 PRACTICE

Use money to show place value.

1. 57¢ 5 tens 7 ones	**2.** 62¢ _____ tens _____ ones
3. 32¢ _____ tens _____ ones	**4.** 84¢ _____ tens _____ ones
5. 17¢ _____ ten _____ ones	**6.** 93¢ _____ tens _____ ones
7. 22¢ _____ tens _____ ones	**8.** 79¢ _____ tens _____ ones
9. 50¢ _____ tens _____ ones	**10.** 41¢ _____ tens _____ one
11. 37¢ _____ tens _____ ones	**12.** 82¢ _____ tens _____ ones
13. 14¢ _____ ten _____ ones	**14.** 72¢ _____ tens _____ ones
15. 49¢ _____ tens _____ ones	**16.** 55¢ _____ tens _____ ones

What comes next in each pattern?

17. 20¢, 25¢, 30¢, _____¢, _____¢, _____¢, _____¢

18. 43¢, 44¢, 45¢, _____¢, _____¢, _____¢, _____¢

Name ______________________________

Quarters and Half Dollars

P 7-4 PRACTICE

Count each group of coins.
Write the total amount.

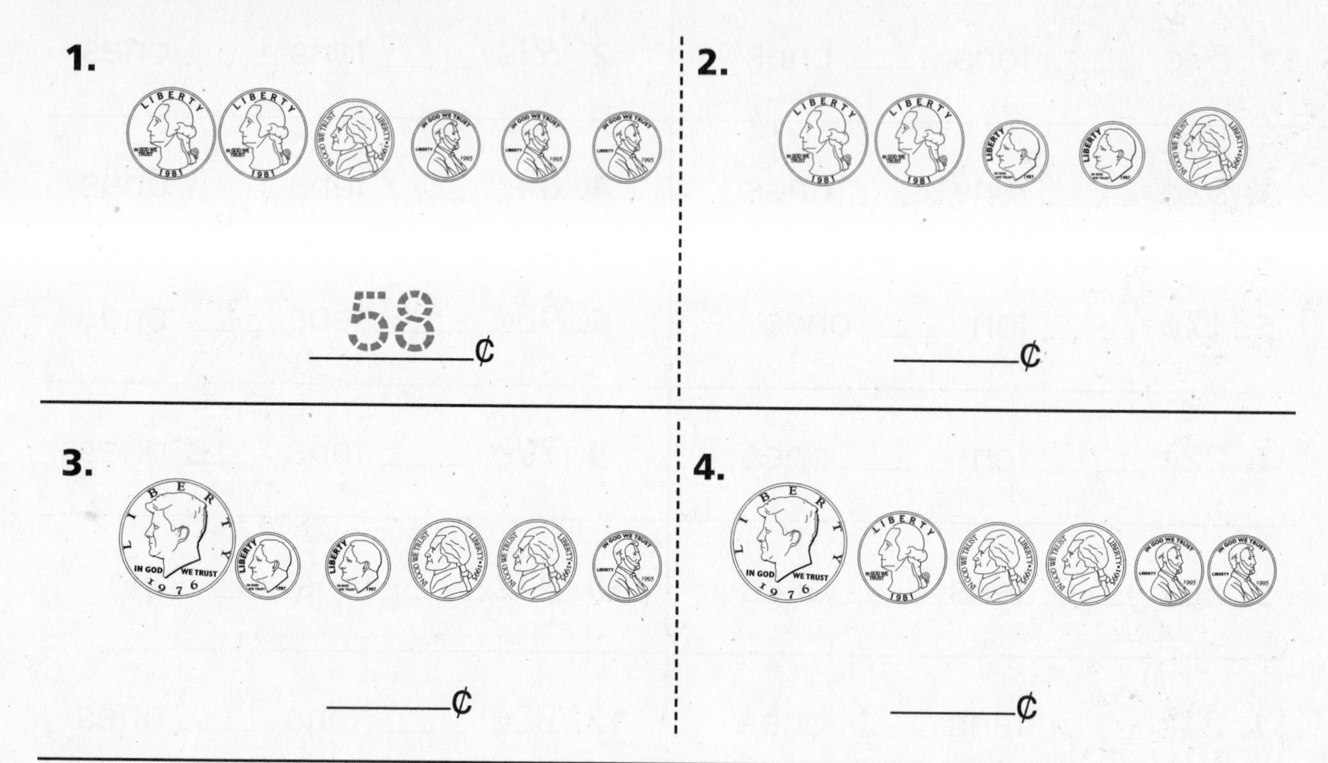

1. 58 ¢

2. ______ ¢

3. ______ ¢

4. ______ ¢

Circle the coins for each amount.

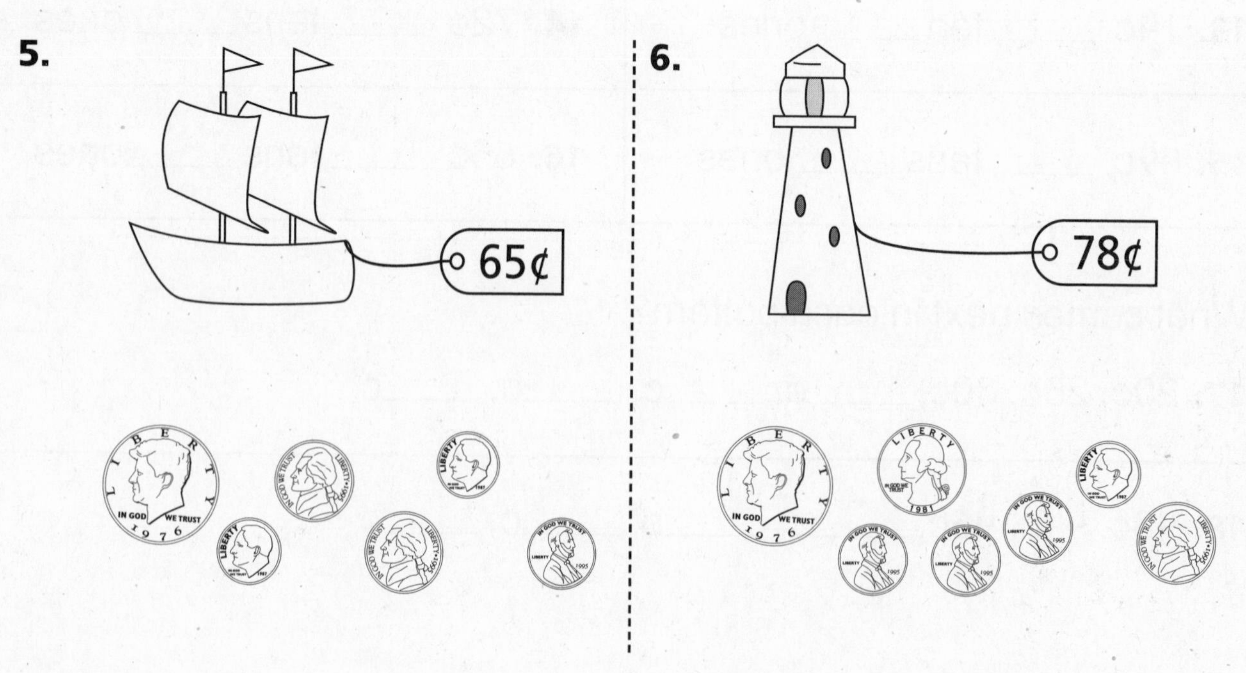

5.

6.

Name ______________________________

Make Equal Amounts

Show the same money amount in two ways.
Circle the way that shows the fewest coins.

Name ____________________

Problem Solving Skill: Reading for Math

Cause and Effect

Kayla wants to buy some stickers.

She helps her neighbor.

She waters the garden.

She gets the mail.

Her neighbor gives Kayla 2 quarters, 1 dime, and 1 nickel.

Choose the best answer. Fill in the ◯.

1. What job does Kayla do?

Ⓐ Buy stickers.

Ⓑ Water the garden.

Ⓒ Mail letters.

Ⓓ Count money.

3. What does Kayla get for helping?

Ⓐ 2 quarters and 3 dimes

Ⓑ 2 quarters, 1 dime, and 1 penny

Ⓒ 2 quarters, 1 dime, and 1 nickel

Ⓓ 2 stickers

2. Why is Kayla helping her neighbor?

Ⓕ Her neighbor likes stickers.

Ⓖ Her mother told her to do that.

Ⓗ She wants to buy stickers.

4. How much money does Kayla get?

Ⓕ 70¢

Ⓖ 60¢

Ⓗ 40¢

Ⓘ 65¢

Name ______________________________

Dollar

Write how much.
Circle the coins that make one dollar.

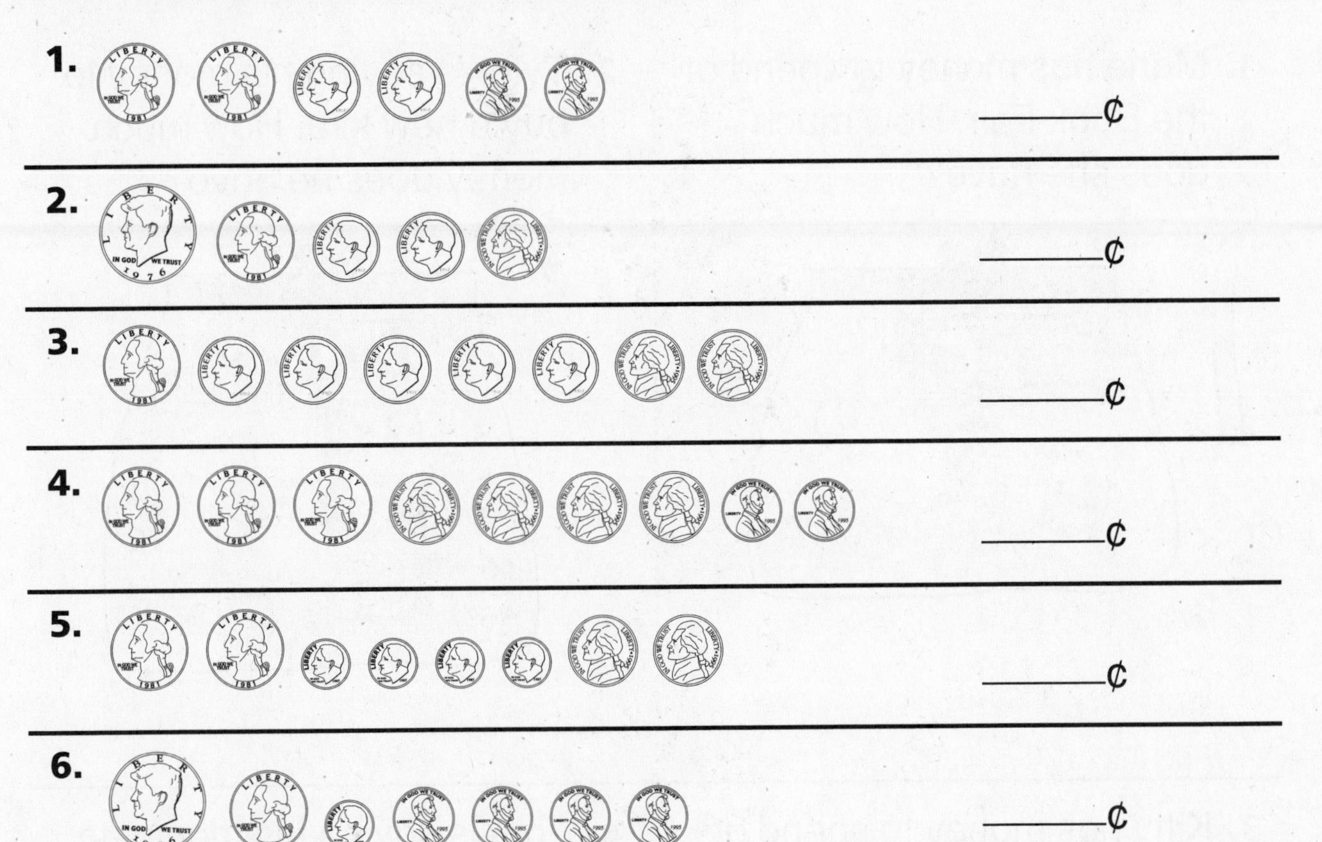

1. ______¢

2. ______¢

3. ______¢

4. ______¢

5. ______¢

6. ______¢

Problem Solving

Solve.

7. Use at least one of each coin to make one dollar. Draw the coins.

Name ______________________________

Dollars and Cents

Count the money.
Write the total amount.

1. Marie has money to spend at the Book Fair. How much does she have?

$ 1 . 54

2. Tyler is saving his money to buy a new kite. How much money does he have?

$____.____

3. Kitty has money to spend at the Game Day. How much does she have to spend?

$____.____

4. Sam is saving his money to buy some new blocks. How much does he have?

$____.____

Name ____

Compare Money Amounts

Count. Is there enough money to buy each item?
Circle yes or no.

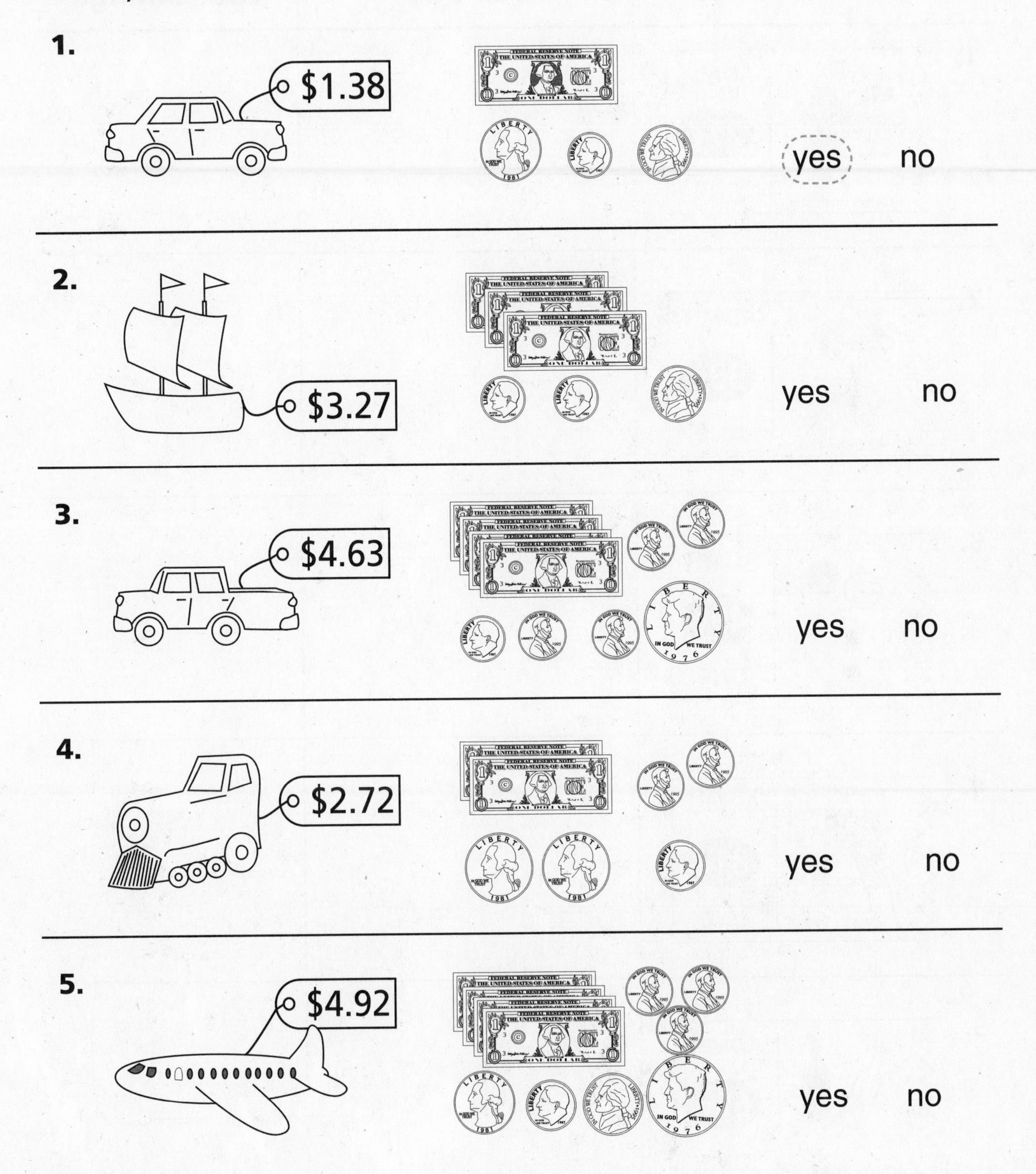

Name ____________________

Make Change

Count up to find the change.

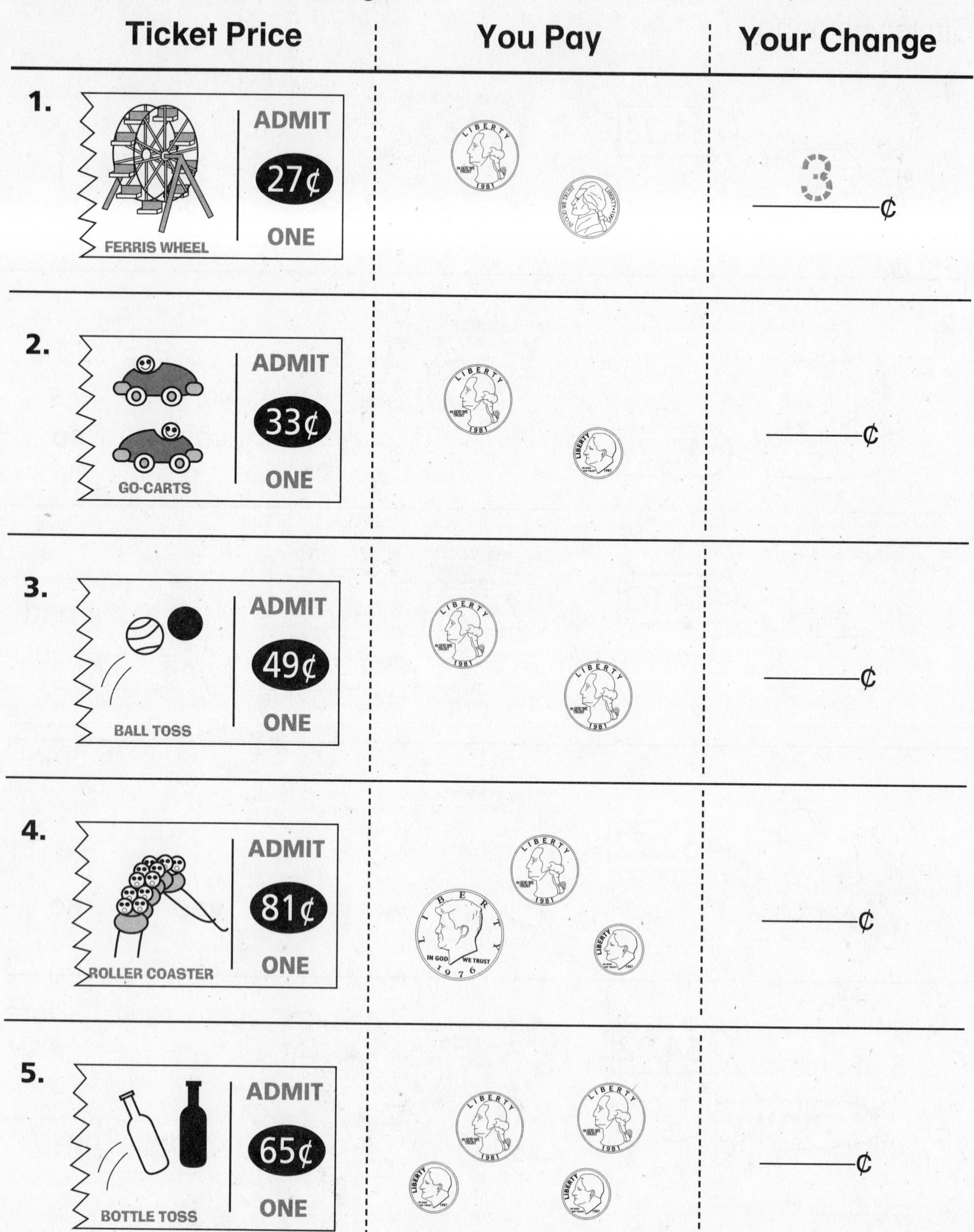

Name ______________________________

Problem Solving: Strategy

Act It Out

Use coins. Act it out.

1. Juan has two quarters and one dime. He bought an apple for 52¢. How much change should he get back?

   8 ¢

2. Rosa has 1 quarter, 2 dimes, and 1 nickel. She bought a pin that costs 46¢. How much change should she get back?

   ______ ¢

3. Mai has 1 quarter, 1 dime, and 1 nickel. She bought a pencil for 35¢. Which coin can she keep?

   ____

4. Maggie has 2 quarters. She bought a sticker that costs 40¢. How much change should she get back?

   ______ ¢

5. Erik has 1 half dollar, 1 dime, and 1 nickel. He bought 2 bananas for 62¢. How much change should he get back?

   ______ ¢

6. Sanjay has a half dollar and 3 dimes. He bought a bagel for 75¢. How much change should he get back?

   ______ ¢

Mixed Strategy Review

Solve.

7. There are 13 otters swimming in the river. 5 otters are sunning on the rocks. How many otters are there in all?

   ______ otters

8. **Write a problem** that you would act out to solve. Share it with others.

   ____

   ____

Name ____________________

Time to the Hour and Half Hour

P 9-1 PRACTICE

Draw the hands to show each time.

1.

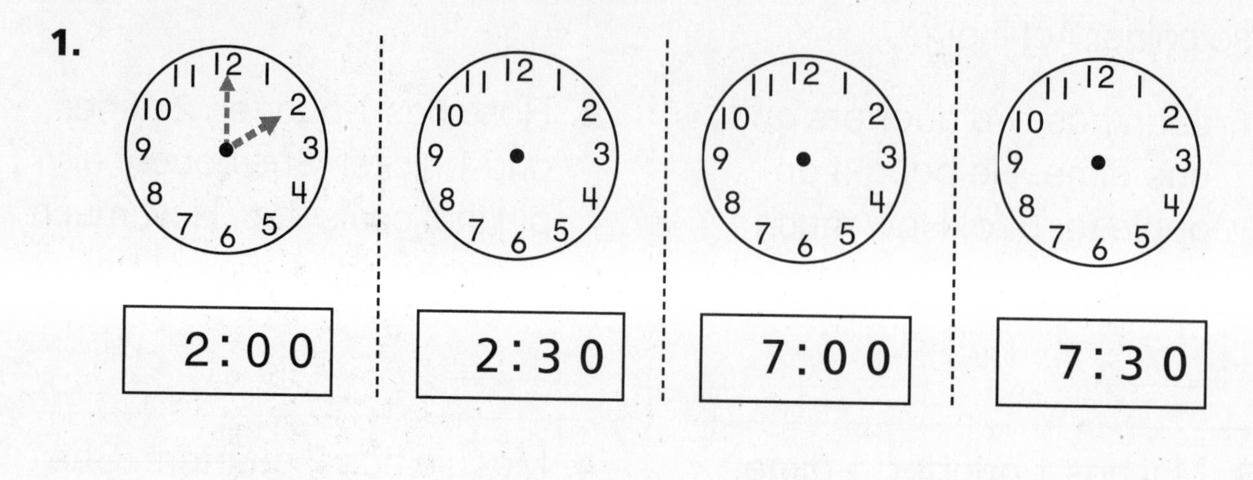

2.

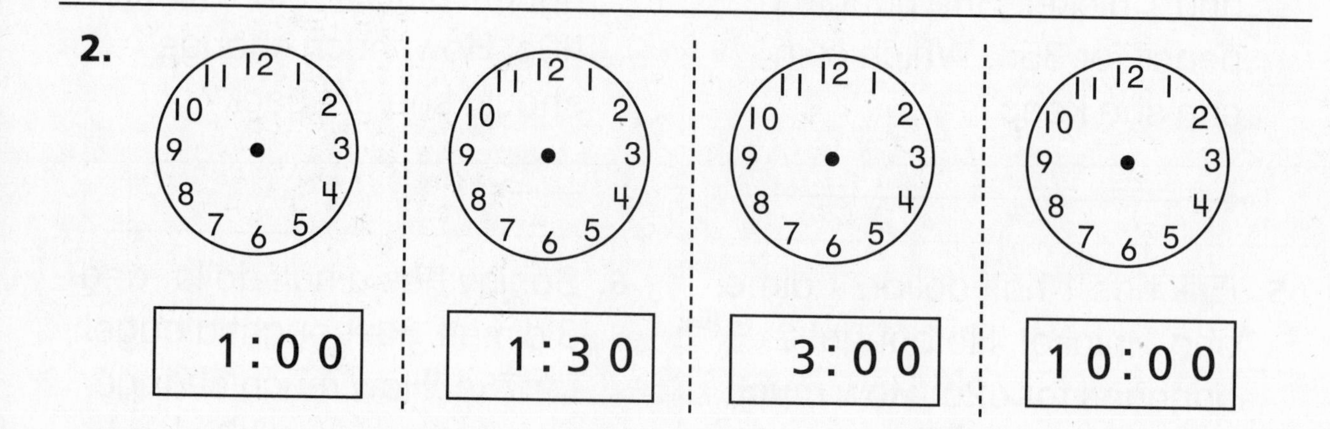

3.

Use with Grade 2, Chapter 9, Lesson 1, pages 155–156.

Name ________________________________

Time to Five Minutes

P 9-2 PRACTICE

Write each time.

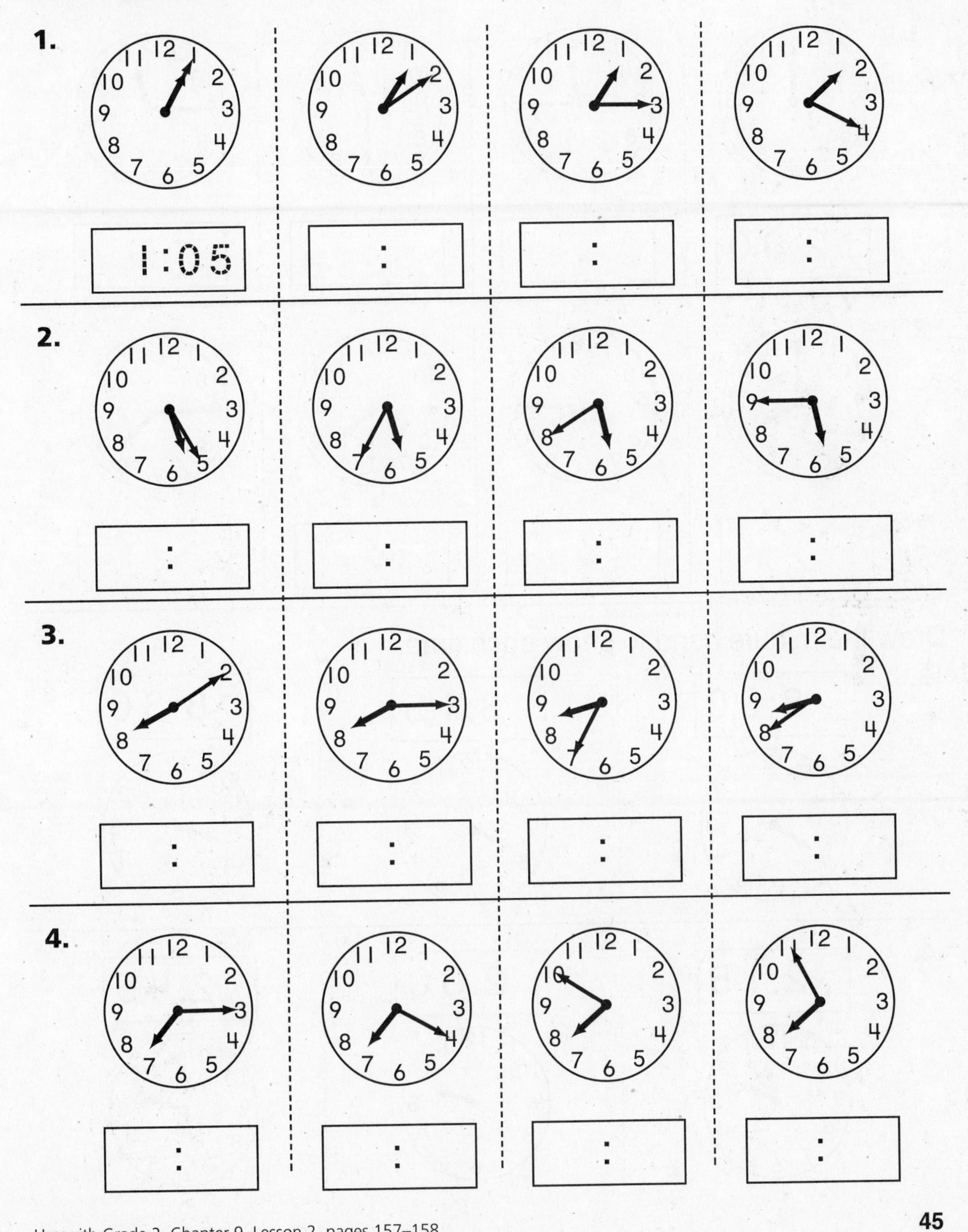

Name ____________________

Time to the Quarter Hour

Write each time.

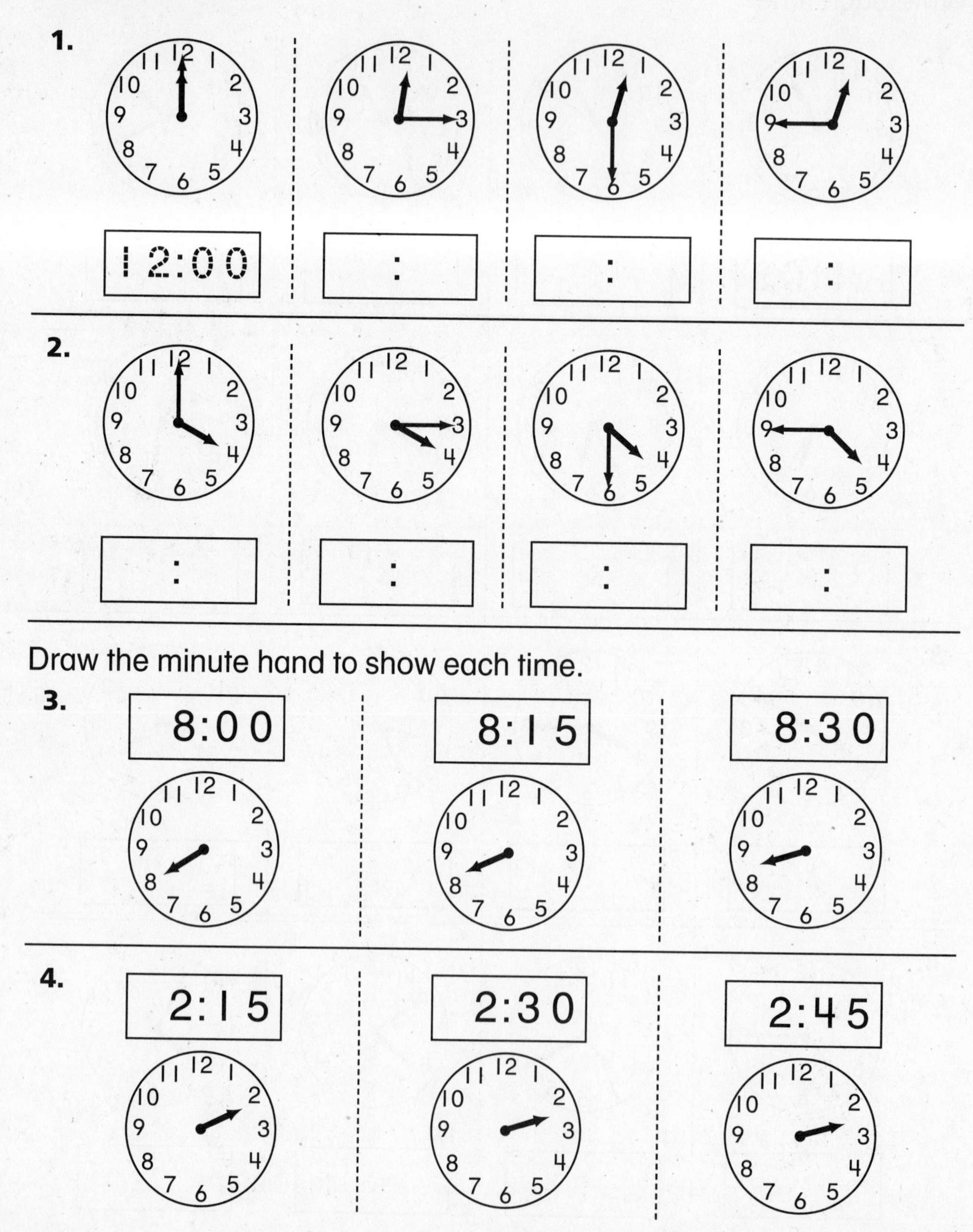

Name ____________________

Time Before and After the Hour

P 9-4 PRACTICE

Write each time more than one way.

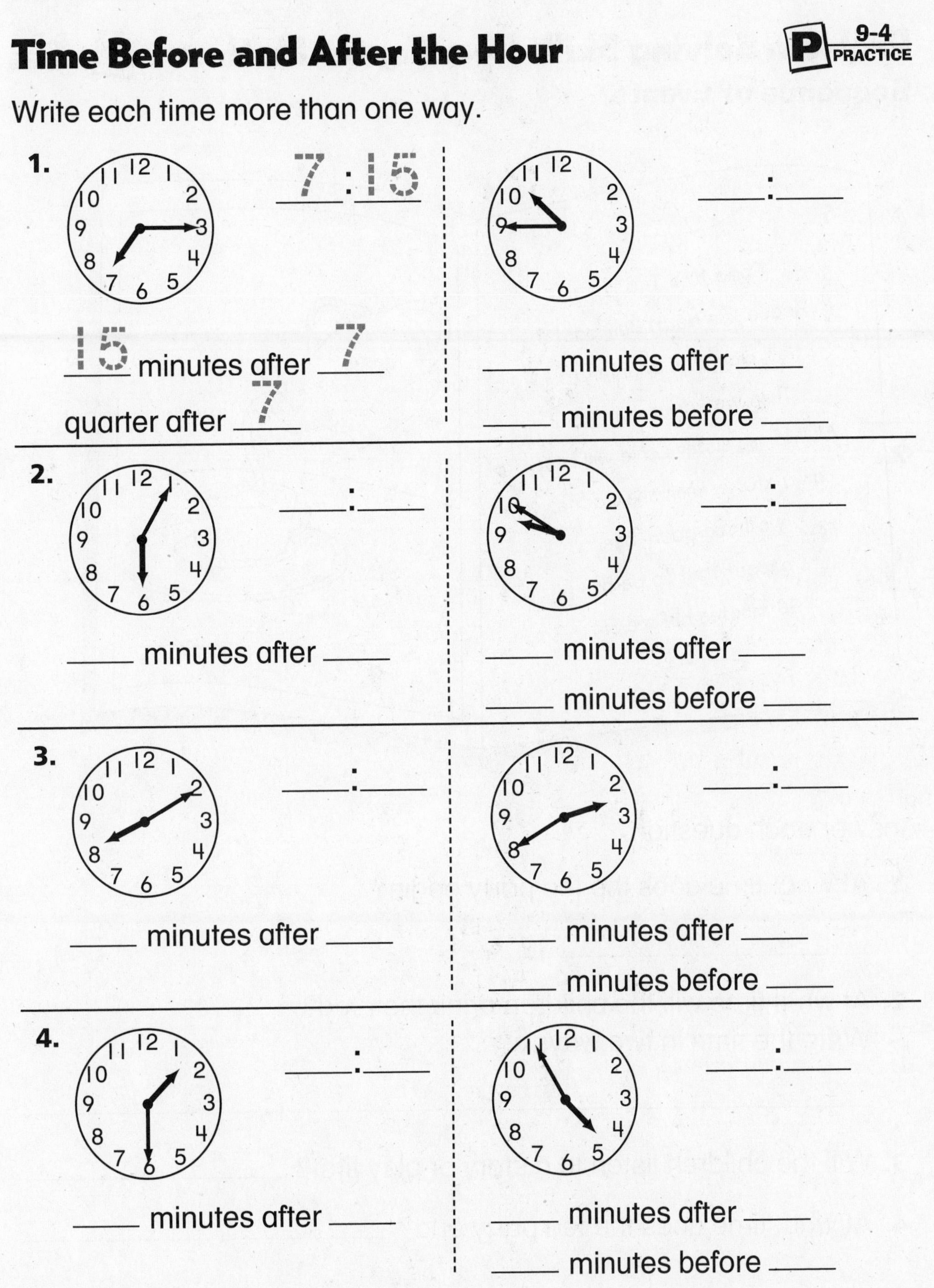

1.

7 : 15

15 minutes after 7

quarter after 7

____ : ____

____ minutes after ____

____ minutes before ____

2.

____ : ____

____ minutes after ____

____ : ____

____ minutes after ____

____ minutes before ____

3.

____ : ____

____ minutes after ____

____ : ____

____ minutes after ____

____ minutes before ____

4.

____ : ____

____ minutes after ____

____ : ____

____ minutes after ____

____ minutes before ____

Name ____________________

Problem Solving Skill: Reading for Math

Sequence of Events

Answer each question.

1. At what time does the tea party begin?

   3:30 ____________________

2. At what time will the children drink their tea? Write the time in two ways.

   ____

3. Will the children listen to a story or play first? ____________________

4. At what time does the tea party end? ____________________

Name ______________________________

A.M. and P.M.

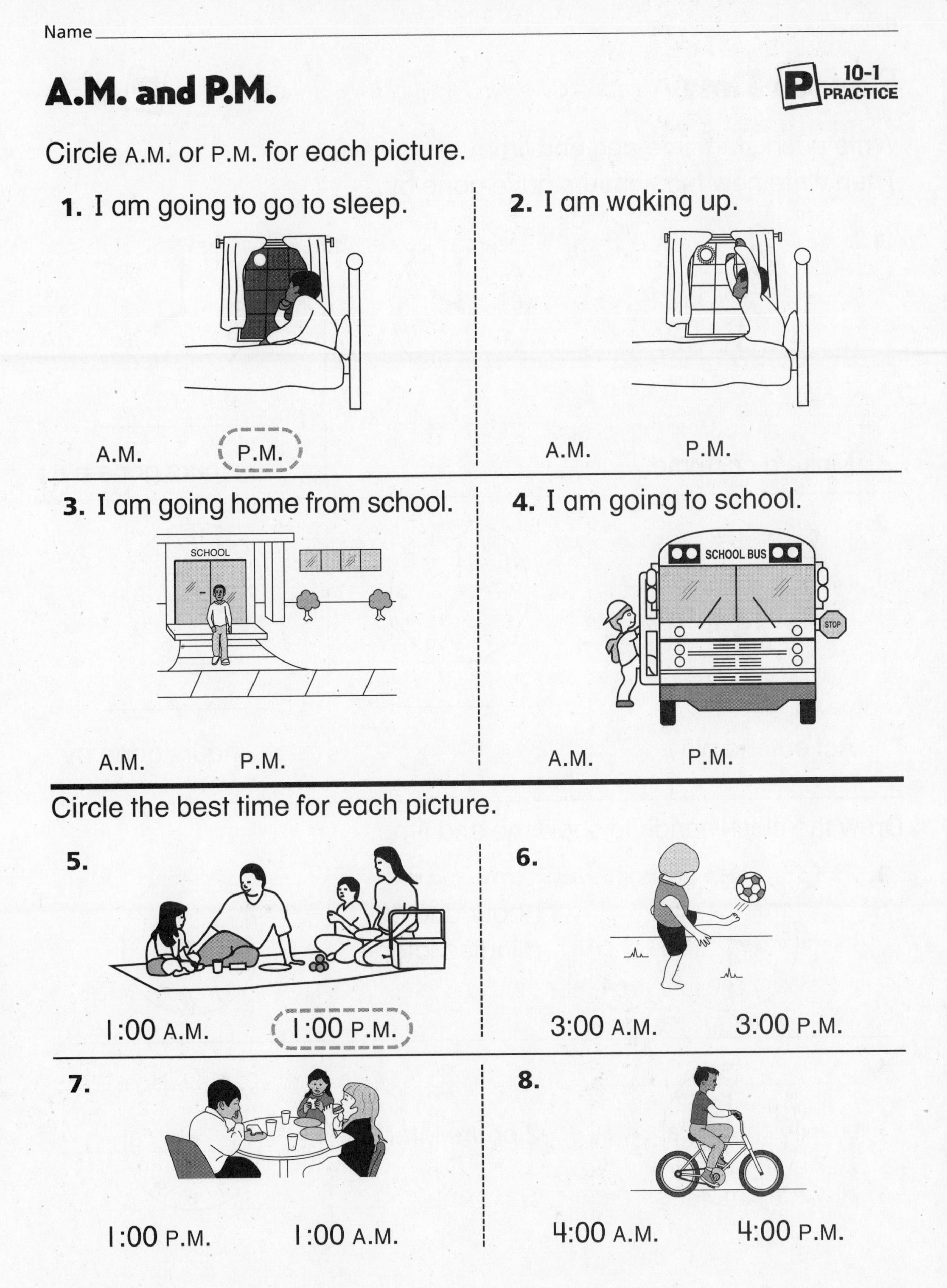

Circle A.M. or P.M. for each picture.

1. I am going to go to sleep.

A.M. P.M.

2. I am waking up.

A.M. P.M.

3. I am going home from school.

A.M. P.M.

4. I am going to school.

A.M. P.M.

Circle the best time for each picture.

5.

1:00 A.M. 1:00 P.M.

6.

3:00 A.M. 3:00 P.M.

7.

1:00 P.M. 1:00 A.M.

8.

4:00 A.M. 4:00 P.M.

Use with Grade 2, Chapter 10, Lesson 1, pages 171–172.

Name ______________________

Elapsed Time

Write each start time and end time.
Then write how many hours have gone by.

Name ____

Calendar

Use the calendar to answer each question.

2004

JANUARY

S	M	T	W	T	F	S
				1	2	3
4	5	6	7	8	9	10
11	12	13	14	15	16	17
18	19	20	21	22	23	24
25	26	27	28	29	30	31

FEBRUARY

S	M	T	W	T	F	S
1	2	3	4	5	6	7
8	9	10	11	12	13	14
15	16	17	18	19	20	21
22	23	24	25	26	27	28
29						

MARCH

S	M	T	W	T	F	S
	1	2	3	4	5	6
7	8	9	10	11	12	13
14	15	16	17	18	19	20
21	22	23	24	25	26	27
28	29	30	31			

APRIL

S	M	T	W	T	F	S
				1	2	3
4	5	6	7	8	9	10
11	12	13	14	15	16	17
18	19	20	21	22	23	24
25	26	27	28	29	30	

MAY

S	M	T	W	T	F	S
						1
2	3	4	5	6	7	8
9	10	11	12	13	14	15
16	17	18	19	20	21	22
23	24	25	26	27	28	29
30	31					

JUNE

S	M	T	W	T	F	S
		1	2	3	4	5
6	7	8	9	10	11	12
13	14	15	16	17	18	19
20	21	22	23	24	25	26
27	28	29	30			

1. On what day of the week does June begin? Tuesday

2. How many Sundays are there in January? In February?

____

3. Which month is the shortest? How many days does it have?

____

4. Which months have 30 days?

____

5. The month just after April is __________.

Name ________________

Time Relationships • Algebra

Circle the best unit to measure the time it takes for each event.

1. to play soccer

(hour) week

to wash your face

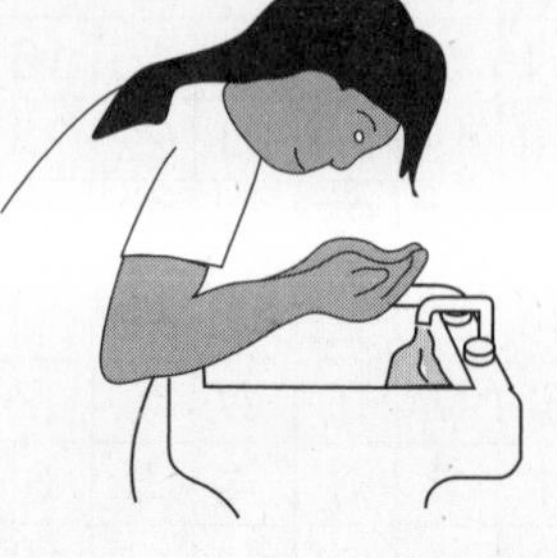

minutes hours

2. to write your name

months minutes

to watch a movie

minute hour

3. to finish second grade

week year

to play a game

minutes days

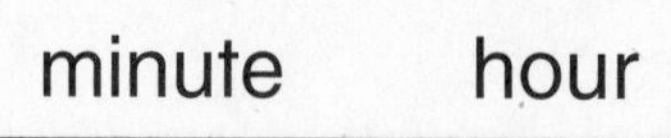

Name ______________________________

Problem Solving: Strategy

Use a Model

January

S	M	T	W	T		S
				1	2	3
4	5	6	7	8	9	10
11	12	13	14	15	16	17
18	19	20	21	22	23	24
25	26	27	28	29	30	31

February

S	M	T	W	T	F	S
1	2	3	4	5	6	7
8	9	10	11	12	13	14
15	16	17	18	19	20	21
22	23	24	25	26	27	28
29						

March

S	M	T	W	T	F	S
	1	2	3	4	5	6
7	8	9	10	11	12	13
14	15	16	17	18	19	20
21	22	23	24	25	26	27
28	29	30	31			

April

S	M	T	W	T	F	S
				1	2	3
4	5	6	7	8	9	10
11	12	13	14	15	16	17
18	19	20	21	22	23	24
25	26	27	28	29	30	

May

S	M	T	W	T	F	S
						1
2	3	4	5	6	7	8
9	10	11	12	13	14	15
16	17	18	19	20	21	22
23	24	25	26	27	28	29
30	31					

June

S	M	T	W	T	F	S
		1	2	3	4	5
6	7	8	9	10	11	12
13	14	15	16	17	18	19
20	21	22	23	24	25	26
27	28	29	30			

Complete.

1. President's Day is the third Monday in February. What is the date?

Monday, February 16

2. Memorial Day is the last Monday in May. What is the date?

Monday, May ______

3. Flag Day is the second Monday in June. What is the date?

Monday, June ______

4. April Fool's Day is the first day in April. What is the date?

__________, April ______

Name ____

Picture and Bar Graphs

P 11-1 PRACTICE

The students voted for their favorite color.

Show the votes on the bar graph.
Use the data.
Color one space for each vote.

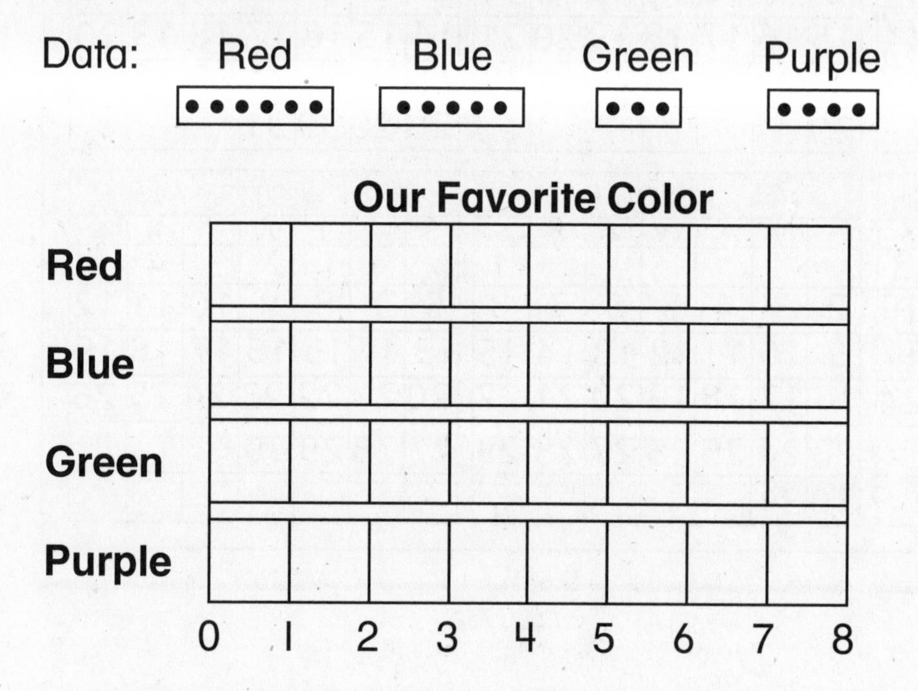

1. Which color is the favorite? ______
2. How many more children chose red than green? ______
3. How many children voted for either green or purple? ______
4. How many children voted in all? ______

Name ______________________________

Surveys

Hannah took a survey.
She asked her friends about their favorite jungle animal.
Use data from the chart to answer each question.

Animal	Votes
elephant	𝍸
lion	𝍸 |||
zebra	||||
monkey	𝍸 𝍸	
giraffe	𝍸 ||||

1. Which animal got 8 votes? lion ______________

2. Which animal got the most votes? ______________

3. Which animal got the fewest votes? ______________

4. Five children said *elephant*. How many more children said *lion*? ______________

5. Four children said *zebra*. How many more children said monkey? ______________

6. How many children in all said *zebra* and *lion*? ______________

7. How many children voted? ______________

Name ______________________________

Make a Bar Graph

Use data from the chart to make a bar graph.

Favorite Bird		
Bird	Tally	Total
Robin	卌 卌 |	11
Blue Jay	卌	
Swan	||||	
Parrot	卌 |||	

Use the totals to complete the bar graph. Color one space for each vote. Then answer each question.

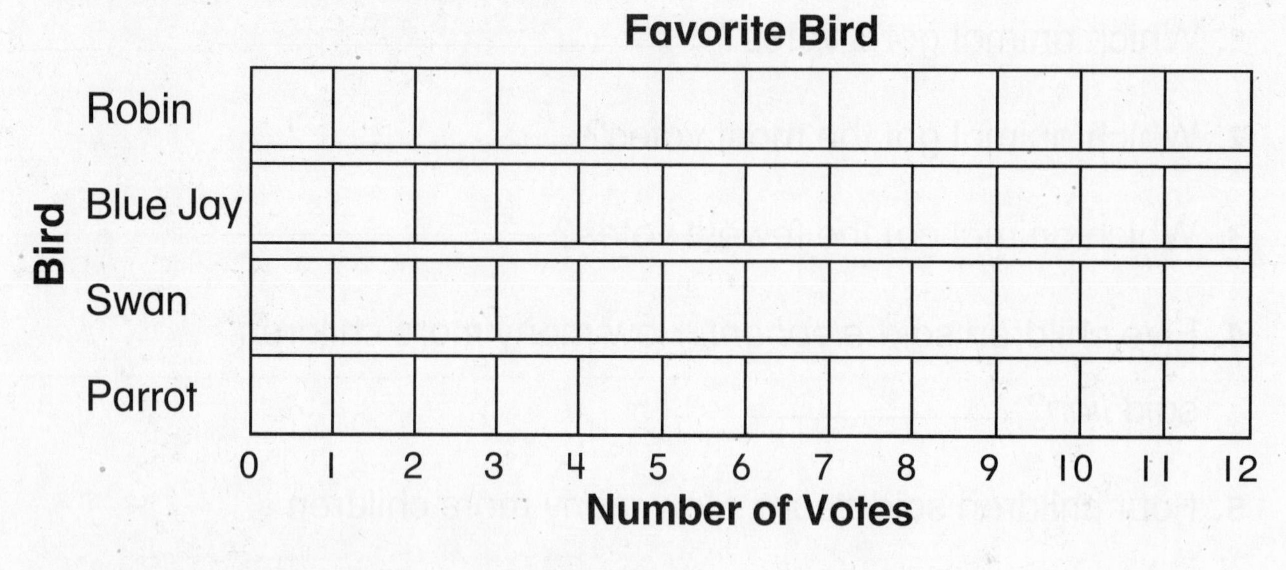

1. Which bird got the most votes? ______________

2. How many more children said *parrot* than said *swan*? ________

3. How many votes for blue jays does the graph show? __________

4. How many children voted in all? __________

Name ____________________

Pictographs

Some children voted for their favorite book.
Use the pictograph. Answer each question.

Favorite Book

Space Raiders	📖📖📖📖📖
Beneath the Sea	📖📖📖📖
House in the Woods	📖📖📖📖📖📖
Puppet Street	📖📖📖📖📖

Key: Each 📖 stands for 2 votes.

1. How many children voted for House in the Woods?
   ____
2. How many more children voted for Puppet Street than voted for Beneath the Sea? ____________
3. How many children in all voted for Space Raiders and Puppet Street? ____________
4. Which book got the fewest number of votes? ____________
5. How many children voted? ____________

Name ________________________________

Line Plots

Use the line plot to answer the questions.

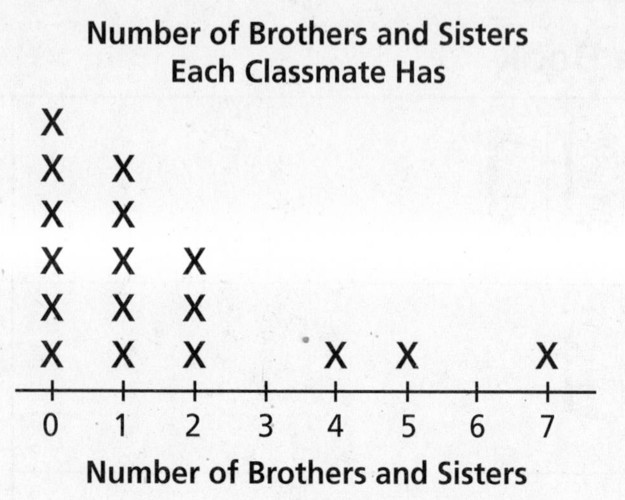

1. How many children have 2 brothers and sisters? ______

2. Which number has the most X's? ______

3. How many children are in the class? ______

4. Does anyone have 7 brothers and sisters? ______

5. How many children have 2 or fewer brothers and sisters? ______

6. How many children have 3 or more brothers and sisters? ______

Name ______________________________

Different Ways to Show Data

Use the tally chart. Make a pictograph and a bar graph to show the data. Then answer the questions.

Our Favorite Food		
Food	Tally	Total
Spaghetti	卌 I	6
Hamburger	IIII	
Pizza	卌 III	

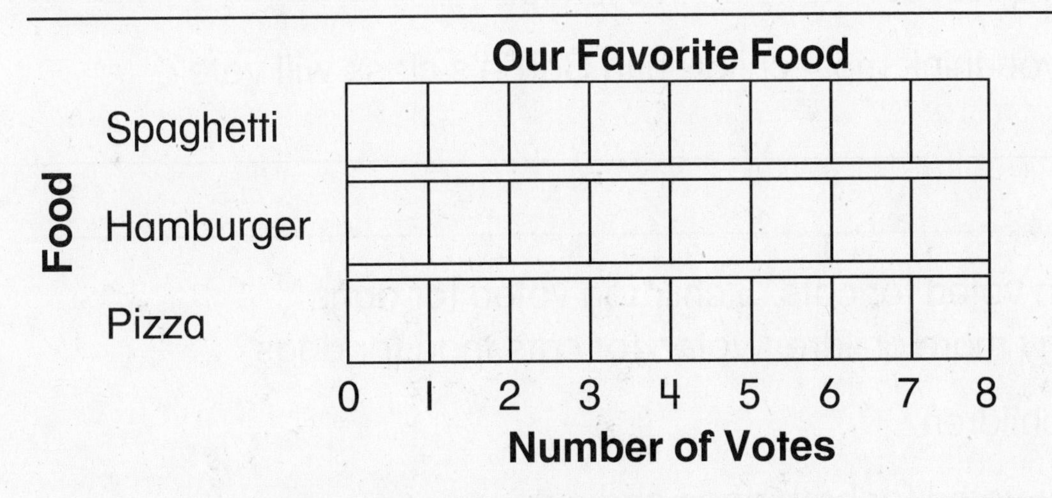

Our Favorite Food

Spaghetti	
Hamburgers	
Pizza	
Key: Each 웃 stands for 2 votes.	

1. Which food got the most votes? ____________

2. Which food got the fewest votes? ____________

3. How many students voted? ________

Name ____________________

Problem Solving Skill: Reading for Math

Make Inferences

Carrie asked her classmates to vote for their favorite pet. John says he likes cats. Kate likes dogs. Maria likes fish.

Carrie's Survey			
Dog	Cat	Fish	None

Answer each question. Make an inference.

1. How do you think most children in Carrie's class will vote?

____

____

2. 7 children voted for cats. 5 children voted for dogs.
How many more children voted for cats than for dogs?

_______ children

3. 4 children voted for both dogs and cats.
3 children voted for none.
How many children in all voted for dogs?

_______ children

4. Use the data from questions 2 and 3.

Did more classmates vote for cats or dogs? ____________

5. Fill in the chart above.

6. Use the data from the chart.

How many children voted? _______ children

Name ______________________________

Explore Regrouping

Use ▢, ▭, and tens/ones mat.

Regroup. Write how many tens and ones.

1. I ten 12 ones

tens	ones
2	2

18 ones

tens	ones

I ten 14 ones

tens	ones

2. I ten 15 ones

tens	ones

I ten 19 ones

tens	ones

I ten 13 ones

tens	ones

3. 12 ones

tens	ones

I ten 16 ones

tens	ones

11 ones

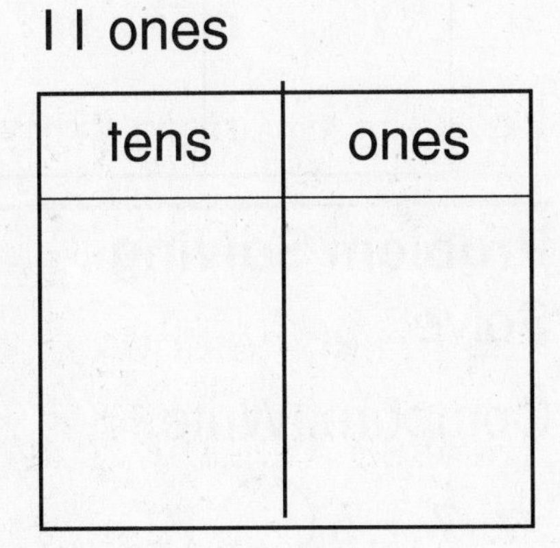

tens	ones

Name ____________________

Addition with Sums to 20

12-2 PRACTICE

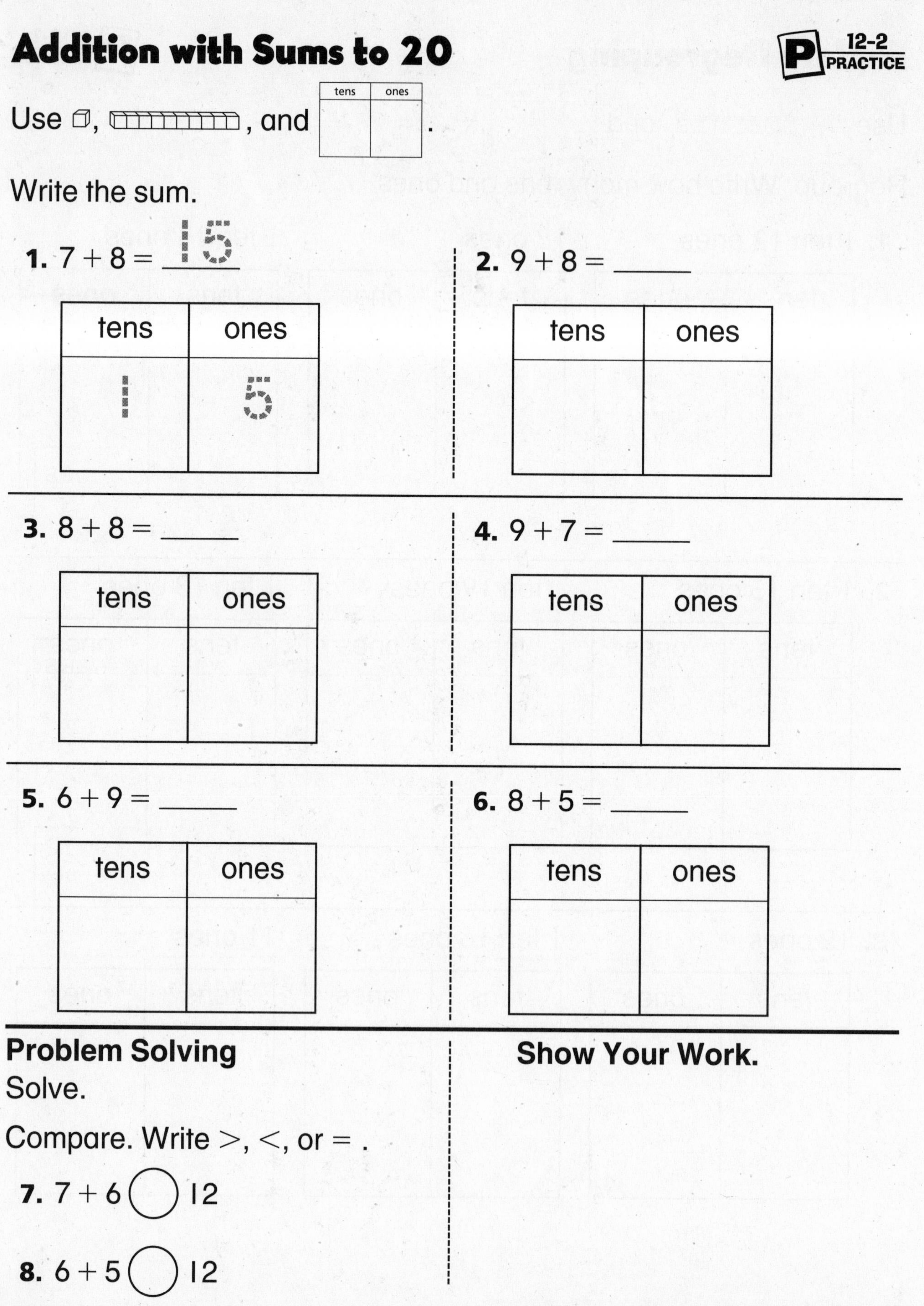

Use [cube], [ten-rod], and [tens | ones mat].

Write the sum.

1. $7 + 8 =$ 15

tens	ones
1	5

2. $9 + 8 =$ ______

tens	ones

3. $8 + 8 =$ ______

tens	ones

4. $9 + 7 =$ ______

tens	ones

5. $6 + 9 =$ ______

tens	ones

6. $8 + 5 =$ ______

tens	ones

Problem Solving

Solve.

Compare. Write >, <, or = .

7. $7 + 6$ ◯ 12

8. $6 + 5$ ◯ 12

Show Your Work.

Name ____________________

Addition with Greater Numbers

12-3 PRACTICE

Add 28 + 17. Use [tens block] or

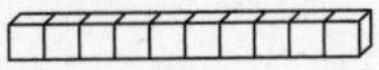

tens	ones

.

Regroup when you have 10 or more ones.
How many tens and ones in all?

45

	Show.		Combine the tens and ones.	How many in all?
1. 28 + 17	tens | ones	tens | ones	tens | ones	tens | ones
2. 14 + 18	tens | ones	tens | ones	tens | ones	tens | ones
3. 25 + 16	tens | ones	tens | ones	tens | ones	tens | ones

Problem Solving

4. How many light bulbs are there in all?

______ light bulbs

Show Your Work

Name ___

Renaming Numbers

12-4 PRACTICE

Write different ways to show each number.

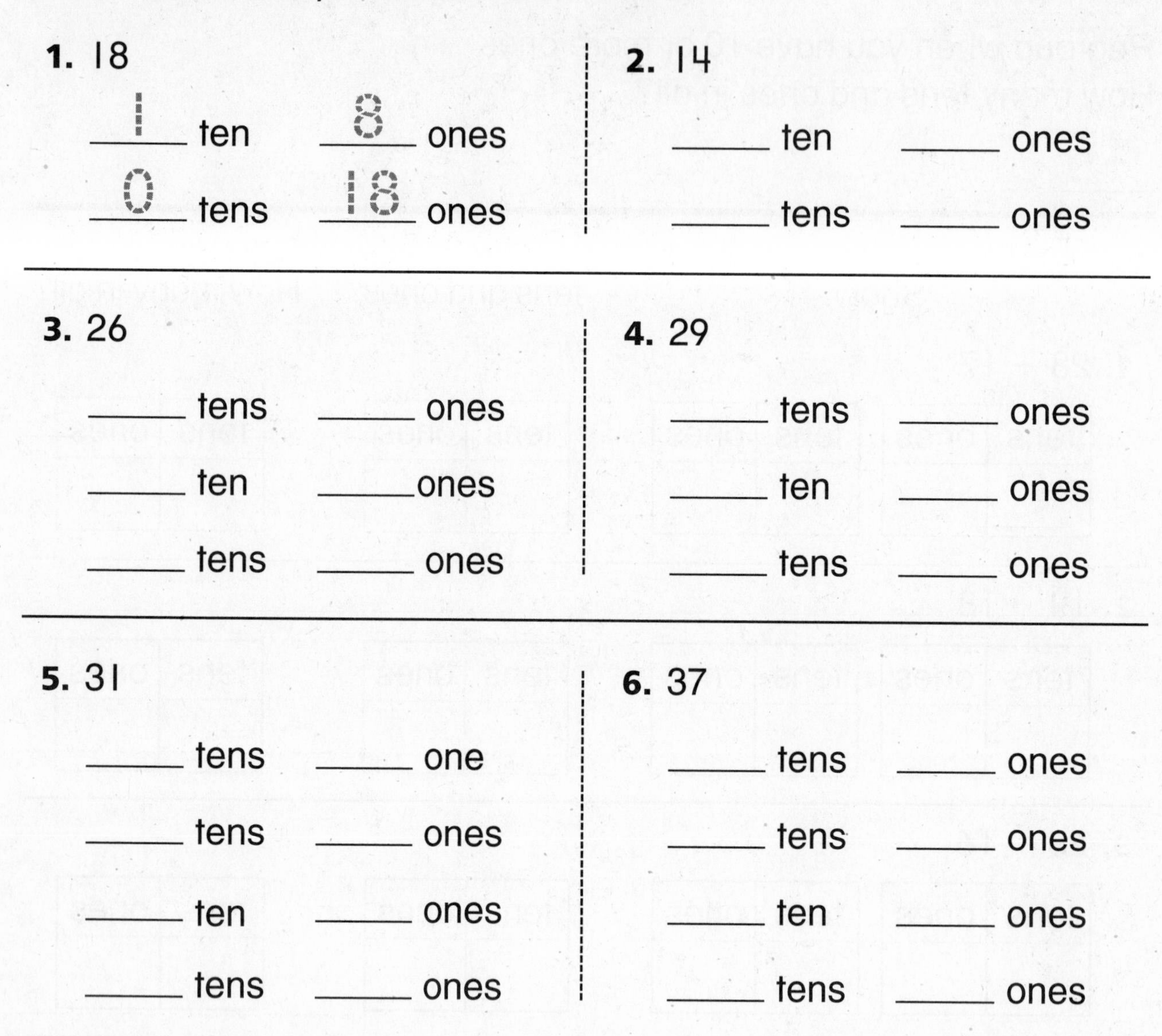

1. 18
   1 ten 8 ones
   0 tens 18 ones

2. 14
   ___ ten ___ ones
   ___ tens ___ ones

3. 26
   ___ tens ___ ones
   ___ ten ___ ones
   ___ tens ___ ones

4. 29
   ___ tens ___ ones
   ___ ten ___ ones
   ___ tens ___ ones

5. 31
   ___ tens ___ one
   ___ tens ___ ones
   ___ ten ___ ones
   ___ tens ___ ones

6. 37
   ___ tens ___ ones
   ___ tens ___ ones
   ___ ten ___ ones
   ___ tens ___ ones

Problem Solving

Solve.

7. Circle all the ways to show 47.

40 + 7	30 + 17	20 + 37

Name ______________________________

Problem Solving: Strategy

Use a Pattern

Use a pattern to solve.

Draw or write to explain.

1. Terry brings 9 cookies to the party. Barb brings 23 cookies. How many cookies do they bring in all?

9 + 23 = 32

9 + 3 = 12
9 + 13 = 22
9 + 23 = 32

2. There are 8 large cartons of fruit juice. There are 36 small cartons of fruit juice. How many cartons in all?

_____ + _____ = _____

3. There are 7 adults. There are 48 children.How many people in all?

_____ + _____ = _____

4. There are 6 brownies left. The chef went into the kitchen and got 56 more. How many brownies are there now?

_____ + _____ = _____

Name ______________________________

Add Tens

Add tens.

1. 6 tens + 3 tens = __9__ tens   3 tens + 2 tens = ______ tens

$$\begin{array}{r} 60 \\ +\ 30 \\ \hline 90 \end{array} \qquad \begin{array}{r} 30 \\ +\ 20 \\ \hline \end{array}$$

2. $\begin{array}{r} 30 \\ +\ 40 \\ \hline \end{array}$ $\begin{array}{r} 50 \\ +\ 20 \\ \hline \end{array}$ $\begin{array}{r} 30 \\ +\ 20 \\ \hline \end{array}$ $\begin{array}{r} 10 \\ +\ 70 \\ \hline \end{array}$

3. $\begin{array}{r} 30 \\ +\ 50 \\ \hline \end{array}$ $\begin{array}{r} 40 \\ +\ 20 \\ \hline \end{array}$ $\begin{array}{r} 20 \\ +\ 60 \\ \hline \end{array}$ $\begin{array}{r} 70 \\ +\ 20 \\ \hline \end{array}$

4. $\begin{array}{r} 50 \\ +\ 10 \\ \hline \end{array}$ $\begin{array}{r} 10 \\ +\ 20 \\ \hline \end{array}$ $\begin{array}{r} 30 \\ +\ 30 \\ \hline \end{array}$ $\begin{array}{r} 40 \\ +\ 50 \\ \hline \end{array}$

5. $\begin{array}{r} 40 \\ +\ 40 \\ \hline \end{array}$ $\begin{array}{r} 20 \\ +\ 50 \\ \hline \end{array}$ $\begin{array}{r} 80 \\ +\ 10 \\ \hline \end{array}$ $\begin{array}{r} 50 \\ +\ 30 \\ \hline \end{array}$

Problem Solving

Solve. Use the number line.

6. What is 30 + 20 + 10? ______

0 10 20 30 40 50 60 70 80 90 100

Use with Grade 2, Chapter 13, Lesson 1, pages 231–232.

Name ______________________________

Count on Tens and Ones to Add

13-2 PRACTICE

Add.

1. $43 + 20 =$ 63 $54 + 7 =$ ____ $35 + 30 =$ ____

2. $18 + 40 =$ 58 $51 + 10 =$ ____ $62 + 4 =$ ____

3. $62 + 10$ $24 + 30$ $40 + 28$ $13 + 70$ $35 + 9$ $55 + 20$

4. $20 + 49$ $34 + 20$ $57 + 8$ $6 + 44$ $80 + 11$ $48 + 40$

5. $35 + 10$ $20 + 53$ $30 + 13$ $44 + 8$ $36 + 50$ $17 + 30$

6. $8 + 23$ $44 + 40$ $38 + 20$ $60 + 18$ $70 + 23$ $57 + 10$

Problem Solving

Solve.

7. There are 30 children in the second grade. There are 45 children in the third grade. How many children are there in all?

____ children

Show Your Work

Name ______________________________

Decide When to Regroup

Add. Use [tens rod] and [ones cube].

	Add.	Do you need to regroup?		How many in all?
1.	27 + 6	yes	no	33
2.	32 + 7	yes	no	
3.	61 + 9	yes	no	
4.	47 + 5	yes	no	
5.	72 + 6	yes	no	
6.	54 + 9	yes	no	
7.	84 + 5	yes	no	
8.	16 + 8	yes	no	

Problem Solving

Solve.

Show Your Work

9. Sam has 93 duck stamps. Lee gives him 4 more. How many stamps does Sam have in all?

_______ stamps

Name ____________________

Add a 1-Digit and a 2-Digit Number

Add. You can use [tens block] and [ones cube] to help.

1.

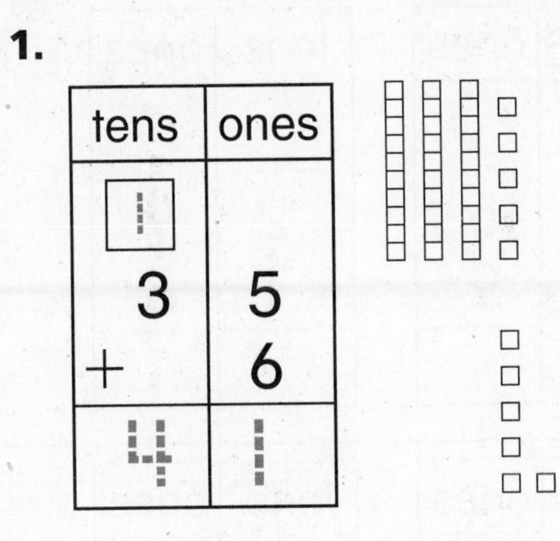

2.

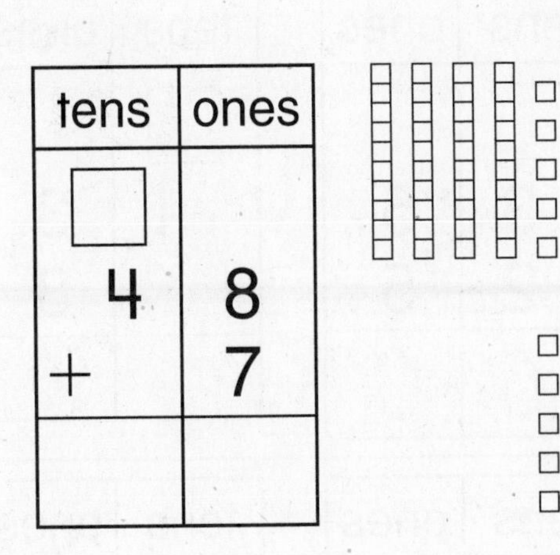

3.

tens	ones
□	
2	3
+	9

4.

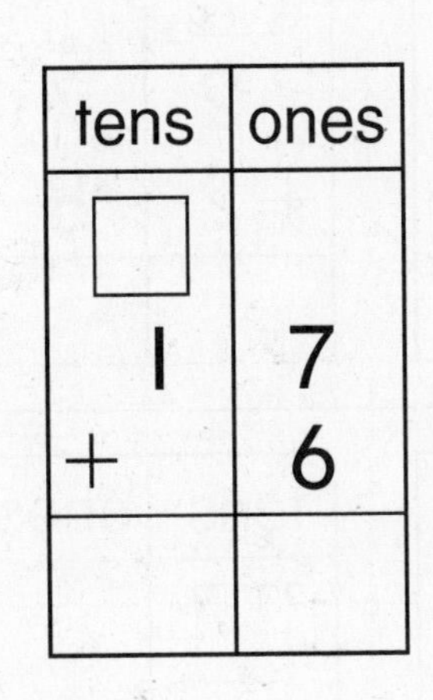

5.

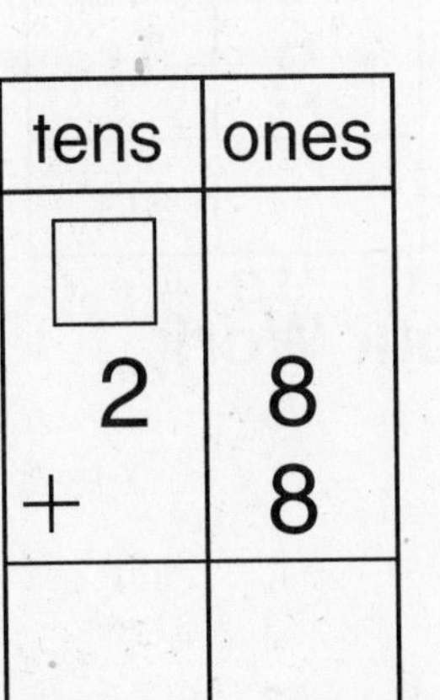

6.

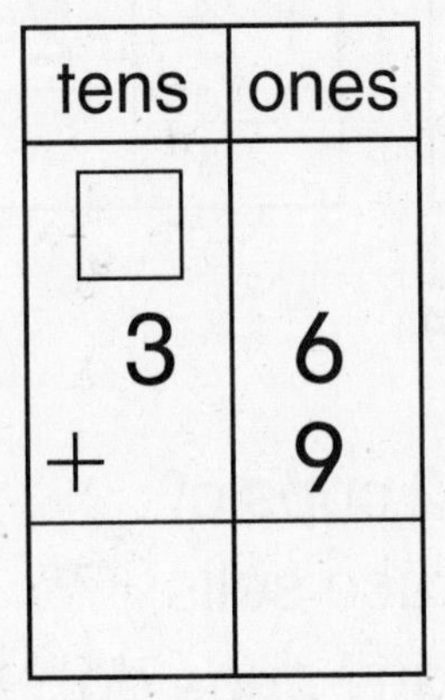

Name ________________________________

Add 2-Digit Numbers

P 13-5 PRACTICE

Add. You can use [tens rod] and [ones cube] to help.

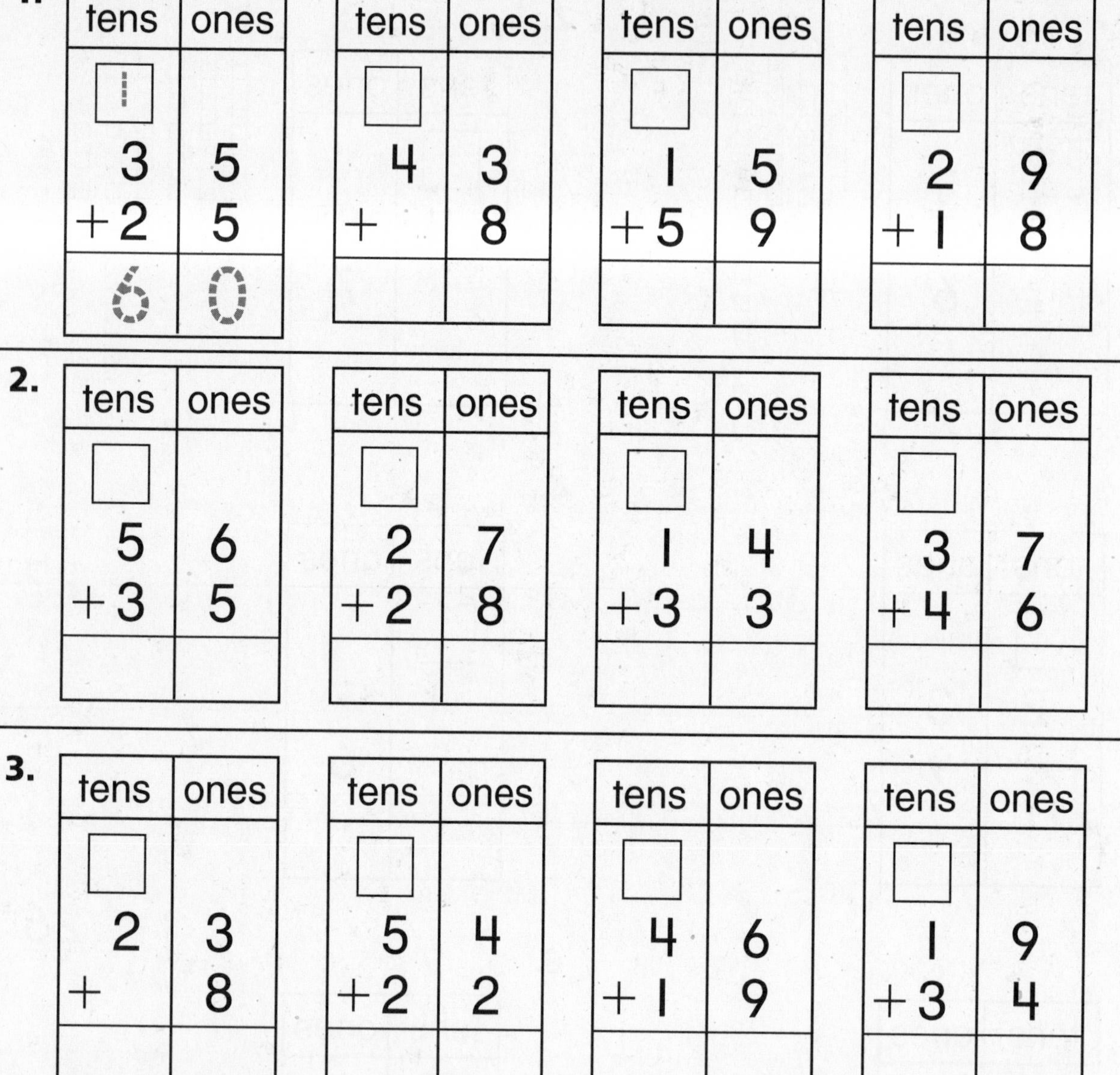

1.

tens	ones	tens	ones	tens	ones	tens	ones
1							
3	5	4	3	1	5	2	9
+2	5	+	8	+5	9	+1	8
6	0						

2.

tens	ones	tens	ones	tens	ones	tens	ones
5	6	2	7	1	4	3	7
+3	5	+2	8	+3	3	+4	6

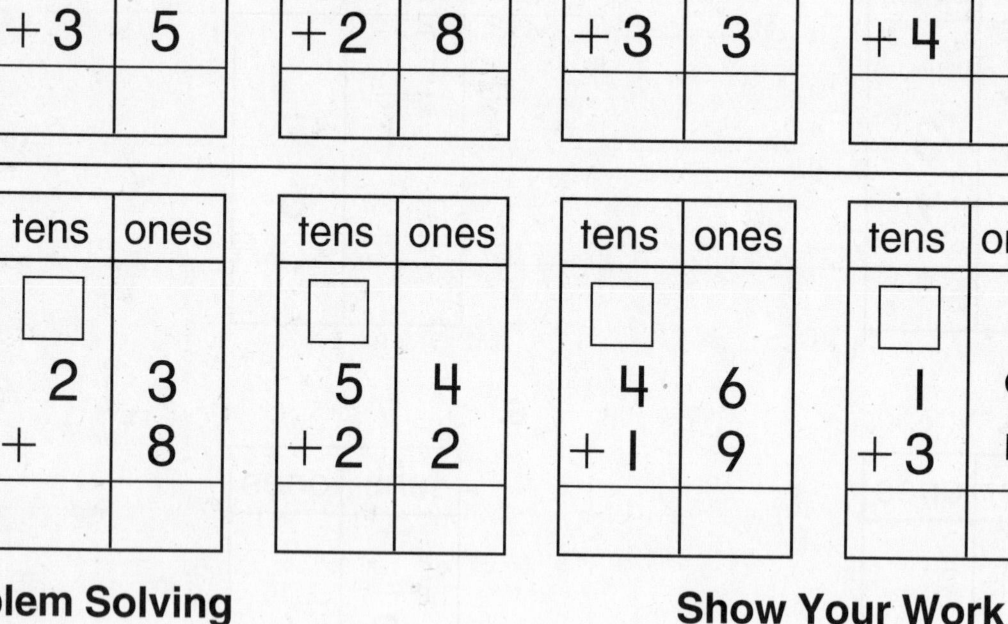

3.

tens	ones	tens	ones	tens	ones	tens	ones
2	3	5	4	4	6	1	9
+	8	+2	2	+1	9	+3	4

Problem Solving

Solve.

Show Your Work

4. Jenny sells 32 cups of lemonade. Bruce sells 17 cups. How many cups do they sell in all?

_______ cups

Use with Grade 2, Chapter 13, Lesson 5, pages 241–242.

Name ______________________________

Practice Addition

Add. You can use [tens block] and [ones cube] to help.

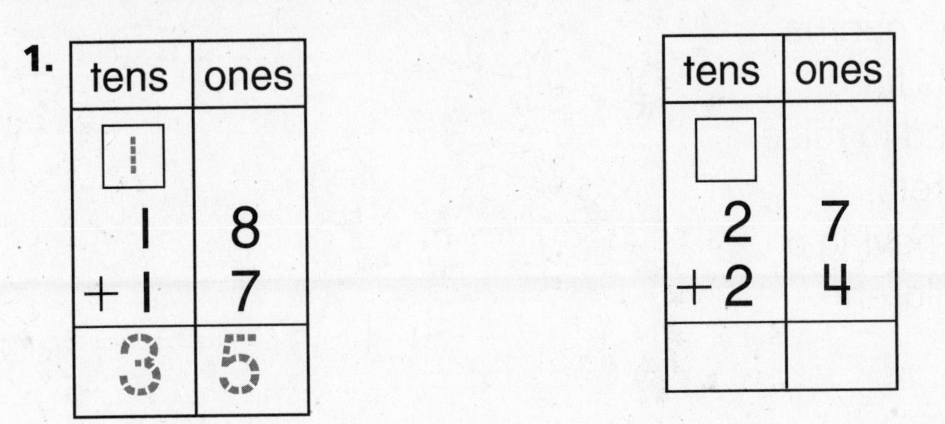

1.

tens	ones
1	
1	8
+ 1	7
3	5

tens	ones
2	7
+ 2	4

2. $25 + 36 = 61$ (regrouped 1)

3. $53 + 9$

4. $38 + 43$

5. $36 + 22$

6. $17 + 35$

7. $74 + 23$

8. $25 + 16$

9. $41 + 26$

10. $55 + 22$

11. $38 + 38$

12. $87 + 6$

13. $45 + 34$

Name ____________________

Problem Solving Skill: Reading For Math

Problem and Solution

The children in the chorus want to buy tickets for a concert. They need to raise money for the tickets. Janelle suggests having a car wash. The children agree.

Answer the questions.

1. What problem do the children have? What do they do to solve the problem?

   ____

   ____

2. The children bought 16 red sponges and 24 blue sponges. How many sponges did they buy in all?

   ________ sponges

3. The children washed 25 cars and 17 minivans. How many vehicles did they wash in all?

   ________ vehicles

4. The children raised 50 dollars washing cars. They raised 34 dollars washing minivans. How much money did they raise in all?

   ________ dollars

Name ________________

Rewrite Addition

14-1 PRACTICE

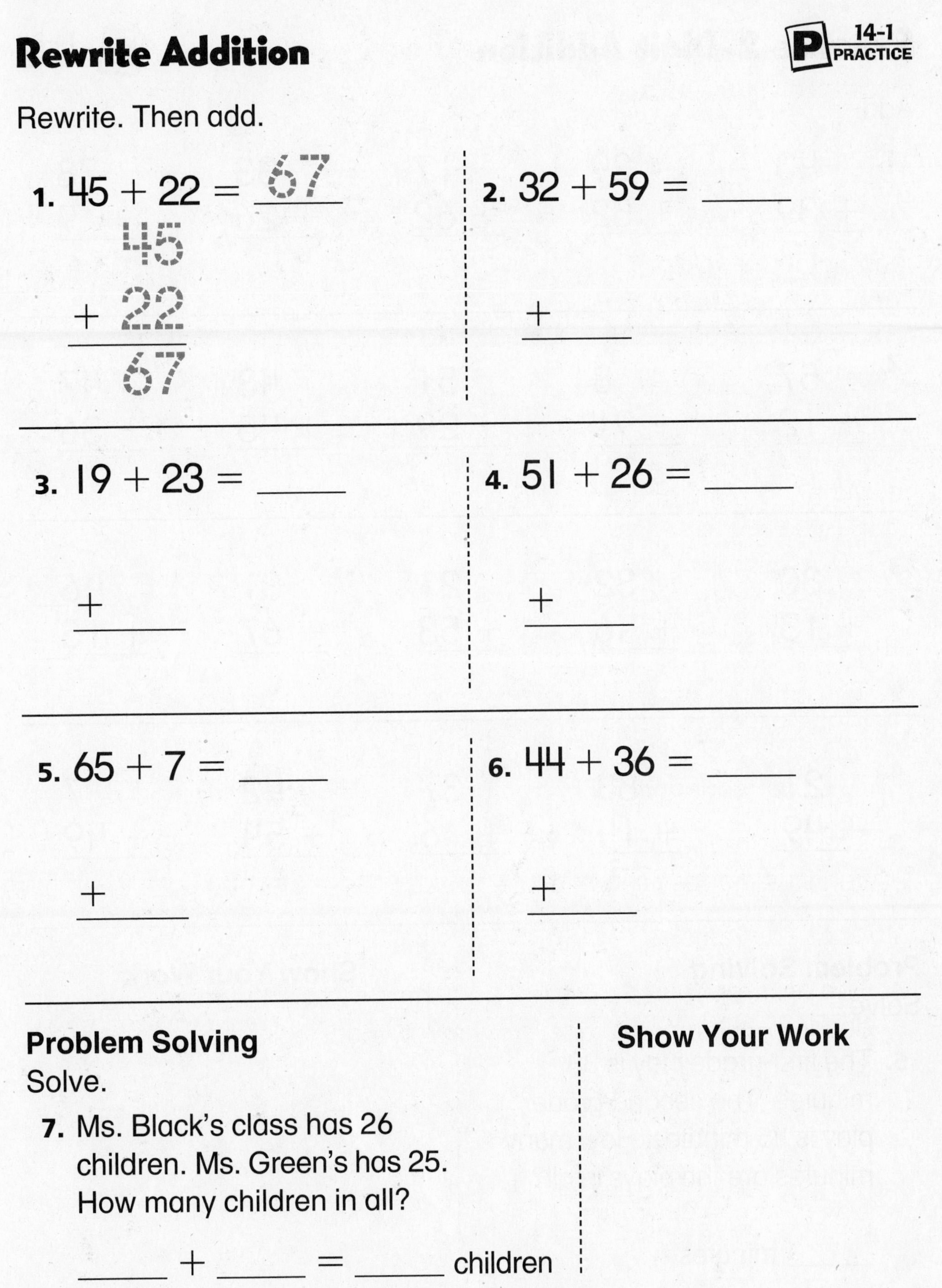

Rewrite. Then add.

1. $45 + 22 =$ 67

$$\begin{array}{r} 45 \\ +\ 22 \\ \hline 67 \end{array}$$

2. $32 + 59 =$ ____

$+$ ____

3. $19 + 23 =$ ____

$+$ ____

4. $51 + 26 =$ ____

$+$ ____

5. $65 + 7 =$ ____

$+$ ____

6. $44 + 36 =$ ____

$+$ ____

Problem Solving

Solve.

7. Ms. Black's class has 26 children. Ms. Green's has 25. How many children in all?

____ + ____ = ____ children

Show Your Work

Name ________________________________

Practice 2-Digit Addition

P 14-2 PRACTICE

Add.

1.

43	29	17	33	78
+ 39	+ 52	+ 62	+ 26	+ 9
82				

2.

57	8	51	43	47
+ 12	+ 91	+ 29	+ 45	+ 36

3.

30	32	21	31	46
+ 18	+ 46	+ 53	+ 67	+ 15

4.

21	83	37	42	7
+ 49	+ 14	+ 6	+ 54	+ 49

Problem Solving

Solve.

5. The first-grade play is 33 minutes. The second-grade play is 45 minutes. How many minutes are the plays in all?

______ minutes

Show Your Work

Name ______________________________

Check Addition • Algebra

14-3 PRACTICE

Add. Check by adding in a different order.

1. $\begin{array}{r} 27 \\ +\ 15 \\ \hline 42 \end{array}$ $\begin{array}{r} 15 \\ +\ 27 \\ \hline 42 \end{array}$

2. $\begin{array}{r} 48 \\ +\ 2 \\ \hline \end{array}$ $\begin{array}{r} \\ +\ \_\_ \\ \hline \end{array}$

3. $\begin{array}{r} 18 \\ +\ 37 \\ \hline \end{array}$ $\begin{array}{r} \\ +\ \_\_ \\ \hline \end{array}$

4. $\begin{array}{r} 62 \\ +\ 9 \\ \hline \end{array}$ $\begin{array}{r} \\ +\ \_\_ \\ \hline \end{array}$

5. $\begin{array}{r} 41 \\ +\ 37 \\ \hline \end{array}$ $\begin{array}{r} \\ +\ \_\_ \\ \hline \end{array}$

6. $\begin{array}{r} 36 \\ +\ 37 \\ \hline \end{array}$ $\begin{array}{r} \\ +\ \_\_ \\ \hline \end{array}$

7. $\begin{array}{r} 52 \\ +\ 26 \\ \hline \end{array}$ $\begin{array}{r} \\ +\ \_\_ \\ \hline \end{array}$

8. $\begin{array}{r} 67 \\ +\ 9 \\ \hline \end{array}$ $\begin{array}{r} \\ +\ \_\_ \\ \hline \end{array}$

9. $\begin{array}{r} 39 \\ +\ 17 \\ \hline \end{array}$ $\begin{array}{r} \\ +\ \_\_ \\ \hline \end{array}$

10. $\begin{array}{r} 53 \\ +\ 8 \\ \hline \end{array}$ $\begin{array}{r} \\ +\ \_\_ \\ \hline \end{array}$

11. $\begin{array}{r} 71 \\ +\ 18 \\ \hline \end{array}$ $\begin{array}{r} \\ +\ \_\_ \\ \hline \end{array}$

12. $\begin{array}{r} 68 \\ +\ 6 \\ \hline \end{array}$ $\begin{array}{r} \\ +\ \_\_ \\ \hline \end{array}$

13. $\begin{array}{r} 69 \\ +\ 21 \\ \hline \end{array}$ $\begin{array}{r} \\ +\ \_\_ \\ \hline \end{array}$

14. $\begin{array}{r} 77 \\ +\ 18 \\ \hline \end{array}$ $\begin{array}{r} \\ +\ \_\_ \\ \hline \end{array}$

15. $\begin{array}{r} 66 \\ +\ 25 \\ \hline \end{array}$ $\begin{array}{r} \\ +\ \_\_ \\ \hline \end{array}$

Problem Solving

Solve.

16. There are 35 girls in the park. There are 47 boys in the park. How many children are in the park?

______ children

Show Your Work

Name ____________________

Estimate Sums

Add. Estimate to see if your answer is reasonable.

1. 54 + 15 = 69 | 50 + 20 = 70

2. 34 + 37 = ____ | + ____

3. 24 + 19 = ____ | + ____

4. 17 + 17 = ____ | + ____

5. 58 + 29 = ____ | + ____

6. 32 + 41 = ____ | + ____

7. 48 + 26 = ____ | + ____

8. 29 + 14 = ____ | + ____

9. 67 + 22 = ____ | + ____

10. 16 + 67 = ____ | + ____

11. 46 + 19 = ____ | + ____

12. 37 + 27 = ____ | + ____

13. 18 + 23 = ____ | + ____

14. 72 + 16 = ____ | + ____

15. 39 + 22 = ____ | + ____

Problem Solving

Solve.

16. There are 34 adults at the Swim Club. There are 57 children at the Swim Club. About how many people are at the Swim Club?

about ______ people

Show Your Work

Name ______________________________

Add Three Numbers

14-5 PRACTICE

Add.

1.	23	41	35	13	25
	14	32	18	24	37
	+ 29	+ 16	+ 25	+ 4	+ 14
	66				

2	8	36	55	11	35
	20	28	13	63	16
	+ 13	+ 32	+ 14	+ 24	+ 34

3.	14	52	44	19	24
	18	20	16	68	3
	+ 14	+ 11	+ 22	+ 12	+ 25

4.	21	37	14	62	43
	18	13	45	11	15
	+ 21	+ 27	+ 3	+ 23	+ 22

Problem Solving

Solve.

Show Your Work

5. There are 34 children in first grade. There are 27 in second grade. There are 31 in third grade. How many children are there in all?

_____ children

Name ________________________________

Problem Solving: Strategy

Choose a Method

Choose a method to solve the problem.
Use mental math, paper and pencil, or a calculator.

Draw or write to explain.

1. The Community Center buys 27 adult tickets and 45 children's tickets for the circus. How many tickets were bought in all?

72 tickets

$$\begin{array}{r} \overset{1}{2}7 \\ +\ 45 \\ \hline 72 \end{array}$$

paper and pencil

2. There are 18 clowns on the stage. Then 22 more clowns come on the stage. How many clowns in all?

______ clowns

3. The soda man sells 36 sodas on Monday. He sells 30 sodas on Tuesday. How many sodas does he sell in all?

______ sodas

4. Laurie saw 18 monkeys during the show. She also saw 12 elephants and 10 seals. How many animals did she see in all?

______ animals

Use with Grade 2, Chapter 14, Lesson 6, pages 263–264.

Name ____________________

Subtract Tens

Subtract tens.

1. 7 tens − 3 tens = __4__ tens   5 tens − 2 tens = ______ tens

$$\begin{array}{r} 70 \\ -\ 30 \\ \hline 40 \end{array} \qquad \begin{array}{r} 50 \\ -\ 20 \\ \hline \end{array}$$

2.

$$\begin{array}{r} 80 \\ -\ 40 \\ \hline \end{array} \qquad \begin{array}{r} 90 \\ -\ 20 \\ \hline \end{array} \qquad \begin{array}{r} 40 \\ -\ 30 \\ \hline \end{array} \qquad \begin{array}{r} 90 \\ -\ 70 \\ \hline \end{array}$$

3.

$$\begin{array}{r} 80 \\ -\ 50 \\ \hline \end{array} \qquad \begin{array}{r} 40 \\ -\ 20 \\ \hline \end{array} \qquad \begin{array}{r} 80 \\ -\ 60 \\ \hline \end{array} \qquad \begin{array}{r} 70 \\ -\ 20 \\ \hline \end{array}$$

4.

$$\begin{array}{r} 50 \\ -\ 10 \\ \hline \end{array} \qquad \begin{array}{r} 70 \\ -\ 10 \\ \hline \end{array} \qquad \begin{array}{r} 60 \\ -\ 30 \\ \hline \end{array} \qquad \begin{array}{r} 90 \\ -\ 50 \\ \hline \end{array}$$

5.

$$\begin{array}{r} 90 \\ -\ 80 \\ \hline \end{array} \qquad \begin{array}{r} 70 \\ -\ 50 \\ \hline \end{array} \qquad \begin{array}{r} 80 \\ -\ 10 \\ \hline \end{array} \qquad \begin{array}{r} 50 \\ -\ 30 \\ \hline \end{array}$$

Problem Solving

Solve.

6. Josie has 6 dimes. She spends 3 of the dimes. How much money does Josie have left?

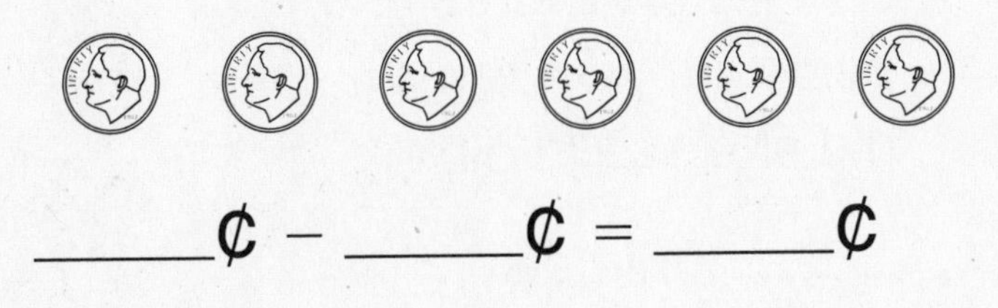

______¢ − ______¢ = ______¢

Name ____________________

Count Back Tens and Ones to Subtract

Count back to subtract.

1.

85	38	57	42	97
− 30	− 6	− 20	− 20	− 4
55				

2.

74	37	86	27	79
− 50	− 30	− 2	− 6	− 40

3.

53	68	43	83	34
− 10	− 5	− 30	− 50	− 3

Find the missing number.

4.

Subtract 2	
55	53
56	____
57	____
58	____

Subtract 4	
55	____
56	____
57	____
58	____

Subtract 30	
55	____
56	____
57	____
58	____

Name ______________________________

Decide When to Regroup

Use models or the tens and ones workmat.

	Subtract.	Do you need to regroup?		How many are left?
1.	54 − 6	(yes)	no	48
2.	32 − 7	yes	no	
3.	82 − 8	yes	no	
4.	47 − 5	yes	no	
5.	63 − 6	yes	no	
6.	91 − 3	yes	no	
7.	32 − 4	yes	no	
8.	79 − 2	yes	no	
9.	54 − 7	yes	no	
10.	38 − 9	yes	no	

Problem Solving

Solve.

11. Sam baked 41 cookies. He ate 4 cookies for his snack. How many cookies are left?

_____ cookies

Show Your Work

Name ____________________

Subtract a 1-Digit Number from a 2-Digit Number

Subtract. Use models or the tens and ones workmat.

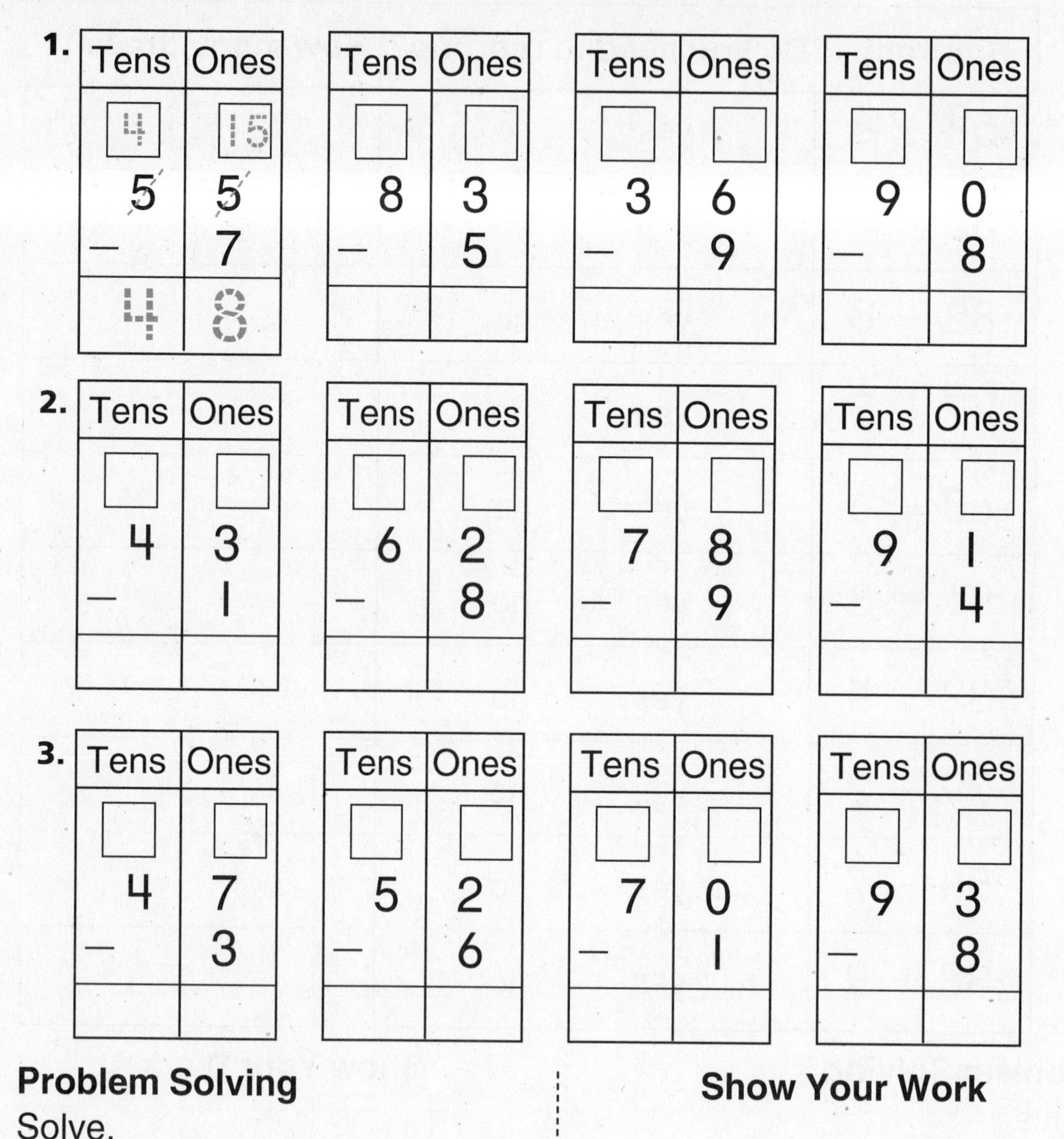

1.

Tens	Ones
4	15
5	5
−	7
4	8

Tens	Ones
8	3
−	5

Tens	Ones
3	6
−	9

Tens	Ones
9	0
−	8

2.

Tens	Ones
4	3
−	1

Tens	Ones
6	2
−	8

Tens	Ones
7	8
−	9

Tens	Ones
9	1
−	4

3.

Tens	Ones
4	7
−	3

Tens	Ones
5	2
−	6

Tens	Ones
7	0
−	1

Tens	Ones
9	3
−	8

Problem Solving

Solve.

4. There are 23 children playing outside. 7 go inside. How many are left outside?

______ children

Show Your Work

Name ____________________

Subtract 2-Digit Numbers

P 15-5 PRACTICE

Subtract. You can use [tens block] and [ones cube] to help.

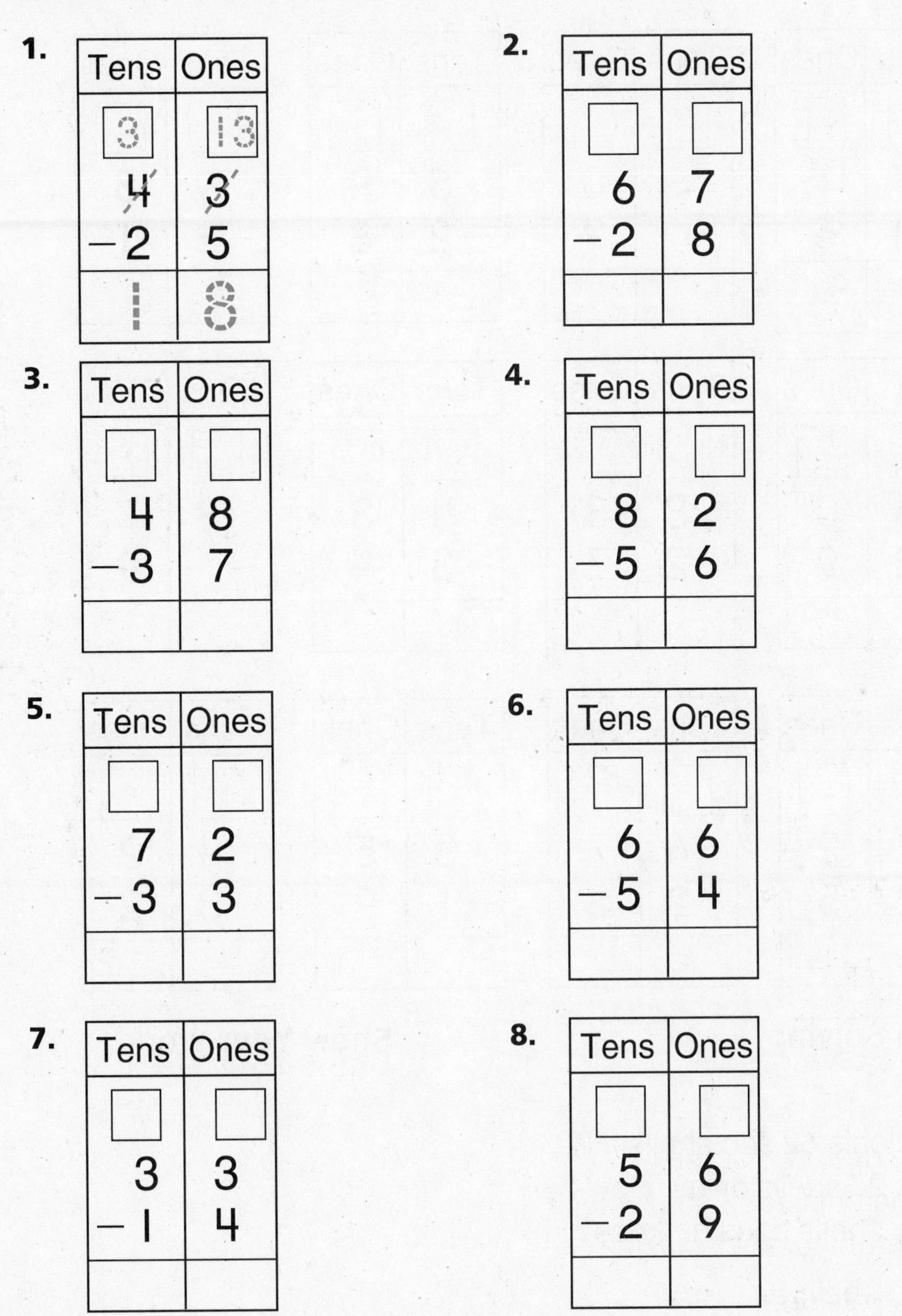

Name ______________________________

Practice Subtraction

Subtract. You can use [tens rod] and [ones cube] to help.

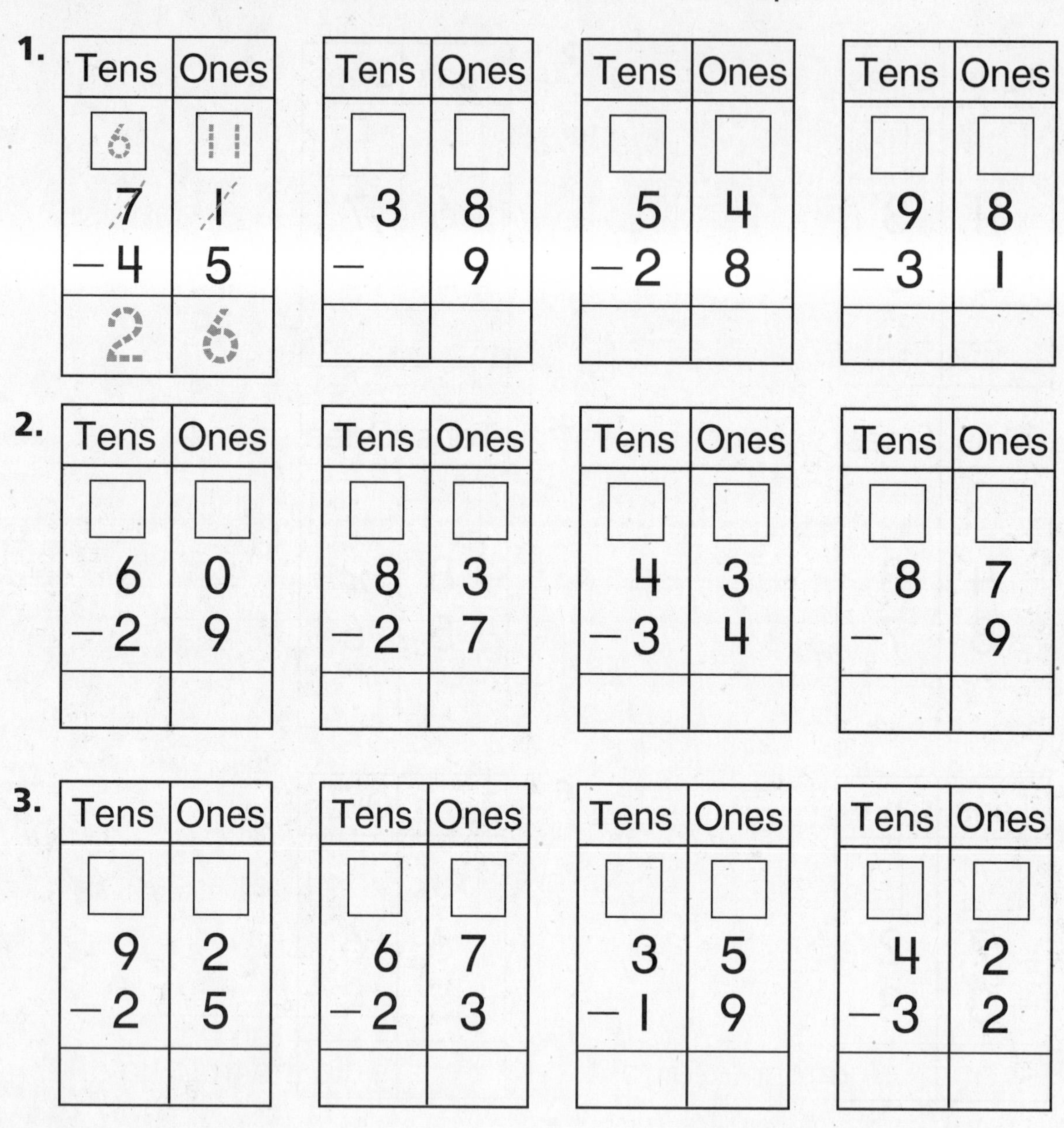

1.

Tens	Ones
6	11
7	1
− 4	5
2	6

Tens	Ones
3	8
−	9

Tens	Ones
5	4
− 2	8

Tens	Ones
9	8
− 3	1

2.

Tens	Ones
6	0
− 2	9

Tens	Ones
8	3
− 2	7

Tens	Ones
4	3
− 3	4

Tens	Ones
8	7
−	9

3.

Tens	Ones
9	2
− 2	5

Tens	Ones
6	7
− 2	3

Tens	Ones
3	5
− 1	9

Tens	Ones
4	2
− 3	2

Problem Solving

Solve.

4. Li collects 62 sea shells. He gives 27 shells away. How many shells does he keep?

______ shells

Show Your Work

Name ______________________________

Problem Solving Skill: Reading for Math

Sequence of Events

A train is headed to the city.
Dover is the first stop. 17 people get on board.
At the next stop, Morris, 15 people get on.
At Madison, 24 people get on.
Then the train goes to the next stop.

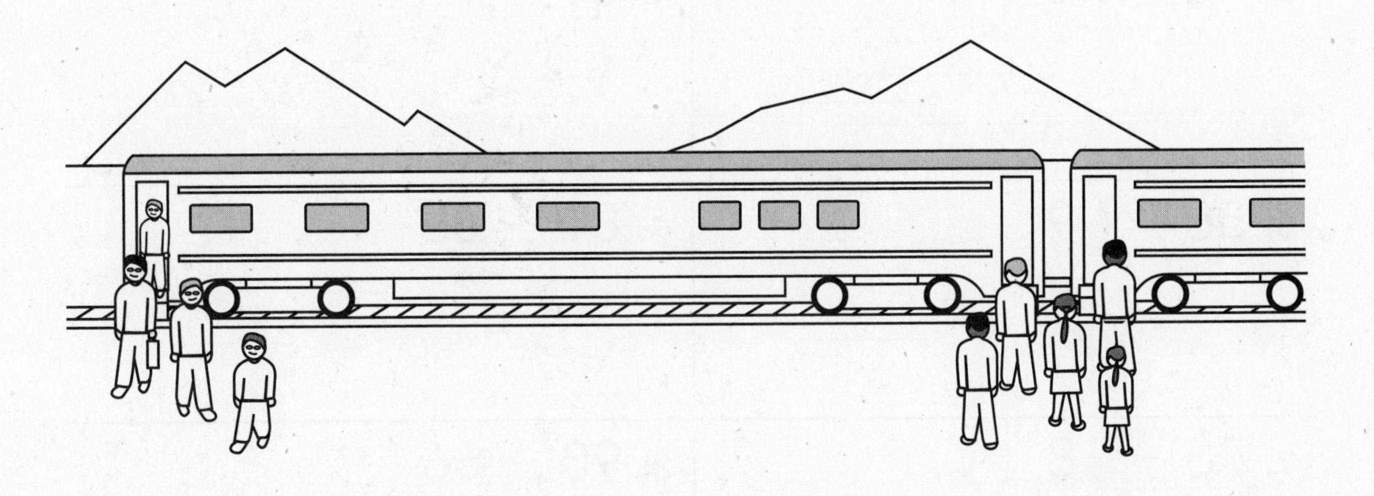

1. In which town did the train stop first?

   ____

2. Did the train stop at Morris **before** or **after** it stopped at Madison?

   ____

3. How many more people got on board in Madison than in Dover? Write a subtraction sentence.

   ____

   _____ people

4. Where did the train go after Madison?

   ____

Name

Rewrite 2-Digit Subtraction

Rewrite. Then subtract.

1. 72 − 24

$$\begin{array}{r} 72 \\ -\ 24 \\ \hline 48 \end{array}$$

2. 67 − 39

3. 46 − 25

4. 51 − 7

5. 58 − 49

6. 81 − 65

7. 73 − 18

8. 90 − 66

Problem Solving

Solve.

9. Last week Mr. Olson ran 45 miles. This week he ran 37 miles. How many more miles did Mr. Olson run last week than this week? Write a number sentence to solve the problem.

_____ − _____ = _____ miles

Show Your Work

Name ______________________

Practice 2-Digit Subtraction

Subtract.

1.	63 − 44 19	85 − 57	71 − 9	90 − 78
2.	35 − 29	83 − 41	57 − 4	38¢ − 25¢
3.	44 − 37	92 − 46	63¢ − 45¢	75 − 37
4.	84 − 36	38 − 7	56 − 51	82 − 45

Problem Solving

Solve.

5. A bottle of water costs 65¢. Patti hands the woman 80¢. How much change should she get back?

______

Show Your Work

Name ______________________________

Check Subtraction

Subtract. Check by adding.

1. $\begin{array}{r} 37 \\ -\ 15 \\ \hline 22 \end{array}$ $\begin{array}{r} 22 \\ +\ 15 \\ \hline 37 \end{array}$ | $\begin{array}{r} 67 \\ -\ 48 \\ \hline \end{array}$ $+$ ____ | $\begin{array}{r} 52 \\ -\ 36 \\ \hline \end{array}$ $+$ ____

2. $\begin{array}{r} 48 \\ -\ 18 \\ \hline \end{array}$ $+$ ____ | $\begin{array}{r} 73 \\ -\ 7 \\ \hline \end{array}$ $+$ ____ | $\begin{array}{r} 82 \\ -\ 68 \\ \hline \end{array}$ $+$ ____

3. $\begin{array}{r} 91 \\ -\ 45 \\ \hline \end{array}$ $+$ ____ | $\begin{array}{r} 35 \\ -\ 17 \\ \hline \end{array}$ $+$ ____ | $\begin{array}{r} 77 \\ -\ 41 \\ \hline \end{array}$ $+$ ____

4. $\begin{array}{r} 56 \\ -\ 28 \\ \hline \end{array}$ $+$ ____ | $\begin{array}{r} 42 \\ -\ 19 \\ \hline \end{array}$ $+$ ____ | $\begin{array}{r} 37 \\ -\ 9 \\ \hline \end{array}$ $+$ ____

5. $\begin{array}{r} 69 \\ -\ 45 \\ \hline \end{array}$ $+$ ____ | $\begin{array}{r} 85 \\ -\ 39 \\ \hline \end{array}$ $+$ ____ | $\begin{array}{r} 62 \\ -\ 23 \\ \hline \end{array}$ $+$ ____

Problem Solving

Solve.

6. There are 46 girls ice skating. There are 67 boys ice skating. How many more boys than girls are ice skating?

______ more boys

Show Your Work

Name ______________________________

Estimate Differences

Subtract. Estimate to see if your answer is reasonable.

1.

74 − 16 = 58	70 − 20 = 50	63 − 21 = ___	− ___	86 − 59 = ___	− ___

2. 54 − 17 = ___   − ___   92 − 26 = ___   − ___   44 − 14 = ___   − ___

3. 76 − 27 = ___   − ___   82 − 37 = ___   − ___   67 − 29 = ___   − ___

4. 38 − 29 = ___   − ___   54 − 19 = ___   − ___   87 − 24 = ___   − ___

5. 64 − 16 = ___   − ___   91 − 73 = ___   − ___   49 − 13 = ___   − ___

Problem Solving

Solve.

Show Your Work

6. A farmer has 72 baskets of apples for sale. She sells 39 baskets of apples. About how many baskets of apples are left?

about ______ baskets

Name ____________________

Mental Math: Strategies

Add or subtract.

1.	$31 + 35 = 66$	$28 + 47$	$36 + 59$	$18 + 43$	$27 + 53$
2.	$55 - 32 = 23$	$73 - 56$	$47 - 34$	$73 - 65$	$90 - 45$
3.	$24 + 57$	$52 + 41$	$13 + 75$	$58 + 6$	$38 + 53$
4.	$74 - 37$	$29 - 16$	$70 - 26$	$97 - 88$	$46 - 8$
5.	$93 - 44$	$56 - 36$	$67 - 39$	$75 - 6$	$31 - 17$

Name ____________________

Problem Solving: Strategy

16-6 PRACTICE

Choose the Operation

Circle add or subtract. Solve.

1. There are 21 children playing tag. Five children are watching. How many more children are playing tag than watching?

   add (subtract)

   21 (–) 5 = 16   16 children

2. Jamie picked 32 strawberries. Alex picked 40 raspberries. How many berries did they pick in all?

   add subtract

   _____ ◯ _____ = _____   _____ berries

3. There are 10 boys and 12 girls in the tap-dance class. How many children are in the tap-dance class?

   add subtract

   _____ ◯ _____ = _____   _____ children

4. Ian has 28 pretzels. He gave all but 6 of them to his sister. How many pretzels did Ian give to his sister?

   add subtract

   _____ ◯ _____ = _____   _____ pretzels

Mixed Strategy Review

Solve.

5. Eve is making up a dance. She leaps, slides, turns, freezes, leaps, slides, turns, freezes, leaps. What move comes next?

   ____

Name ______________________________

Nonstandard Units of Length

17-1 PRACTICE

Estimate. Then use [paper clip] to measure.

Find objects like the ones shown.

1.

Estimate ____ [paper clip] measure ____ [paper clip]

2.

glue

Estimate ____ [paper clip] measure ____ [paper clip]

Problem Solving

Solve.

3. Measure with [cube]. Then measure with [paper clip].

About ____ [cube] about ____ [paper clip]

Are your answers the same or different? Explain why.

____

____

Name ___________________________________

Measure to the Nearest Inch

Find the objects below in your classroom.
Estimate. Then measure and record.
Write how many inches or feet.

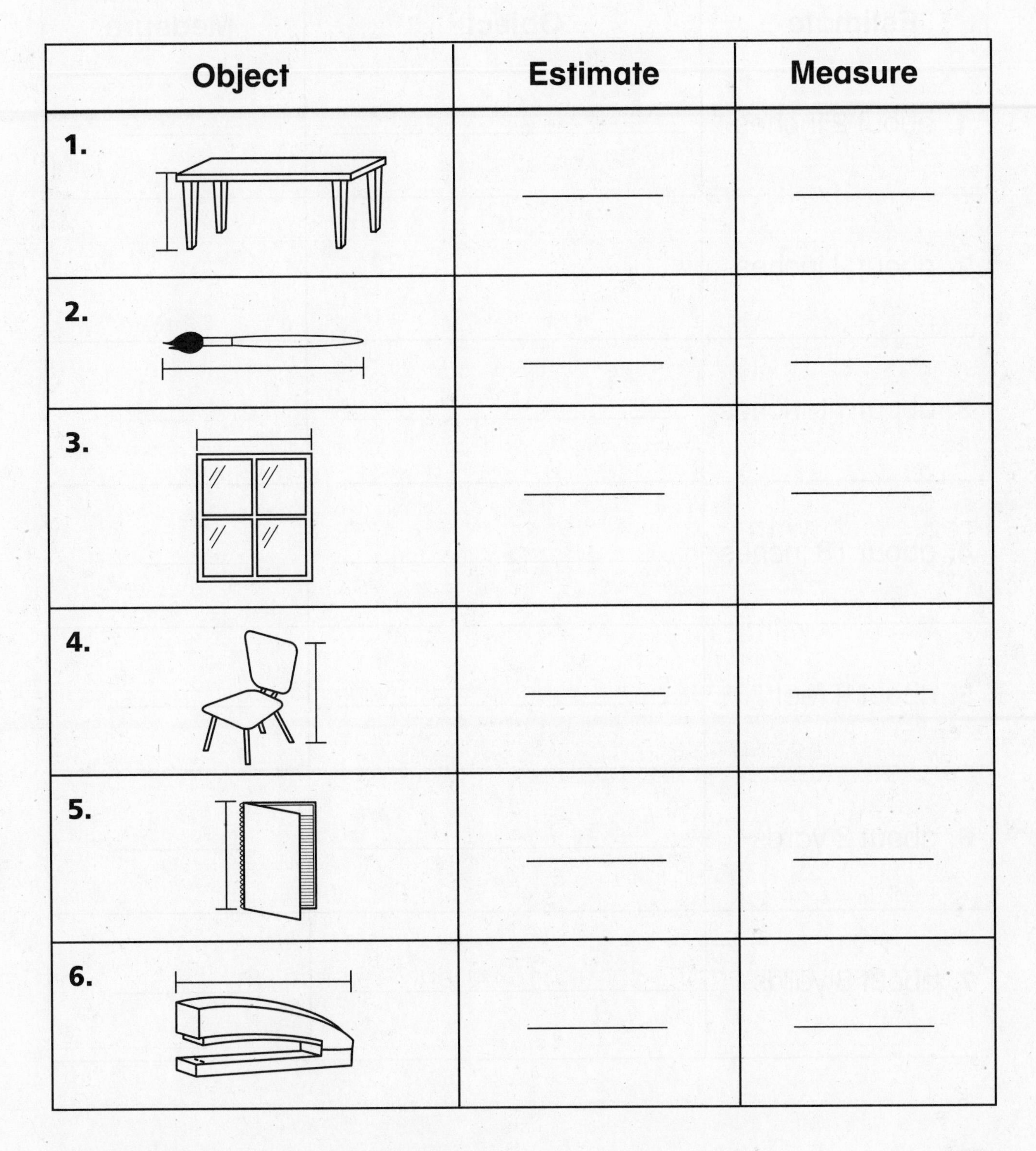

Object	Estimate	Measure
1.	______	______
2.	______	______
3.	______	______
4.	______	______
5.	______	______
6.	______	______

Name ______________________________

Inch, Foot, and Yard

Estimate.
Find an object for each length.

Estimate	Object	Measure
1. about 2 inches	______________	________
2. about 4 inches	______________	________
3. about 11 inches	______________	________
4. about 18 inches	______________	________
5. about 4 feet	______________	________
6. about 2 yards	______________	________
7. about 3 yards	______________	________

Name ______________________________

Centimeter and Meter

Use a centimeter ruler to measure.

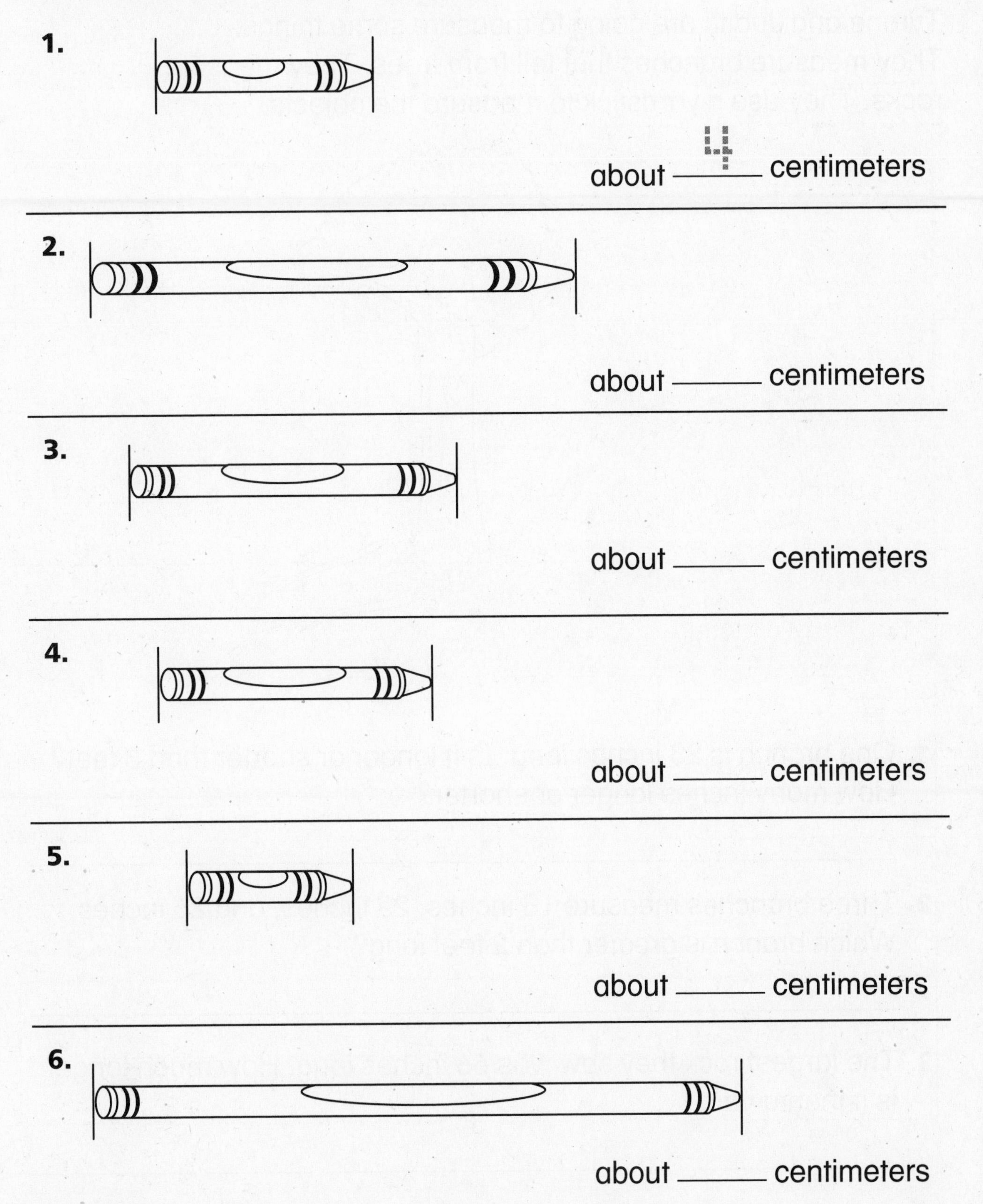

Name ______________________________

Problem Solving Skill: Reading for Math

Compare and Contrast

Tyrone and Judith are going to measure some things. They measure branches that fell from trees. They measure rocks. They use a yardstick to measure the objects.

1. One branch is 28 inches long. Is it longer or shorter than 3 feet? How many inches longer or shorter?

____

2. Three branches measure 18 inches, 22 inches, and 26 inches. Which branch is greater than 2 feet long?

____

3. The largest rock they saw was 66 inches long. How much longer is it than a yard?

____

Name ______________________________

Explore Capacity

Which container holds more milk?

The pitcher will hold about 2 containers of milk.

About how many milk cartons does each container hold?
You can use water and a milk carton to measure.

Container	Estimate	Measure

Name ____________________

Fluid Ounce, Cup, Pint, Quart, and Gallon

Circle the better estimate.

1.

more than 1 cup

less than 1 cup

2.

more than 1 cup

less than 1 cup

3.

more than 1 cup

less than 1 cup

4.

MILK

more than 1 fluid ounce

less than 1 fluid ounce

5.

more than 1 pint

less than 1 pint

6.

more than 1 pint

less than 1 pint

7.

more than 1 quart

less than 1 quart

8.

more than 1 gallon

less than 1 gallon

Use with Grade 2, Chapter 18, Lesson 2, pages 335–336.

Name ____________________

Ounce and Pound

18-3 PRACTICE

Circle the better estimate.

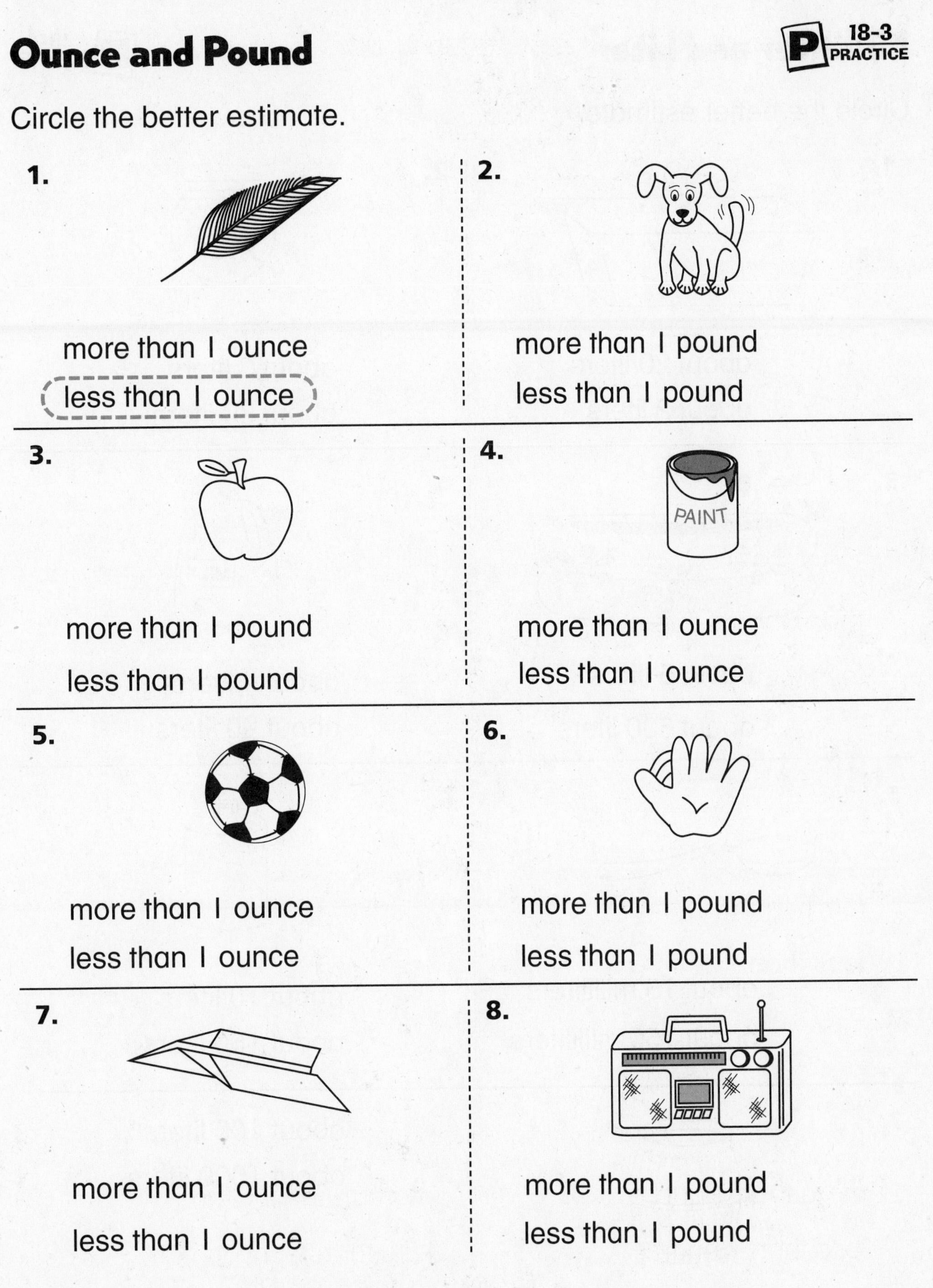

1. more than 1 ounce / less than 1 ounce

2. more than 1 pound / less than 1 pound

3. more than 1 pound / less than 1 pound

4. more than 1 ounce / less than 1 ounce

5. more than 1 ounce / less than 1 ounce

6. more than 1 pound / less than 1 pound

7. more than 1 ounce / less than 1 ounce

8. more than 1 pound / less than 1 pound

Name ____________________

Milliliter and Liter

Circle the better estimate.

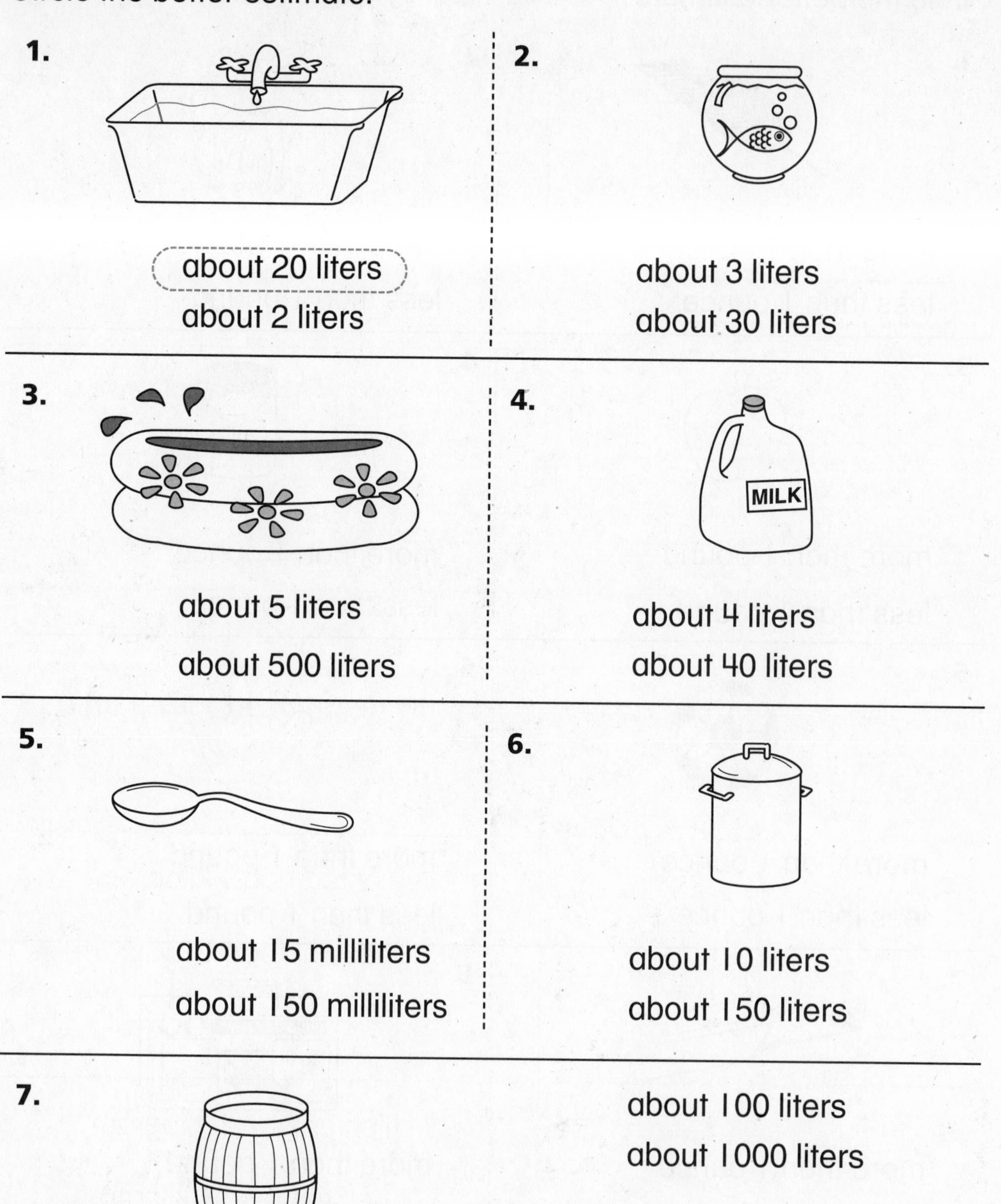

1. about 20 liters
   about 2 liters

2. about 3 liters
   about 30 liters

3. about 5 liters
   about 500 liters

4. about 4 liters
   about 40 liters

5. about 15 milliliters
   about 150 milliliters

6. about 10 liters
   about 150 liters

7. about 100 liters
   about 1000 liters

Name ____________________

Gram and Kilogram

P 18-5 PRACTICE

Think about the objects pictured.
Circle the better estimate.

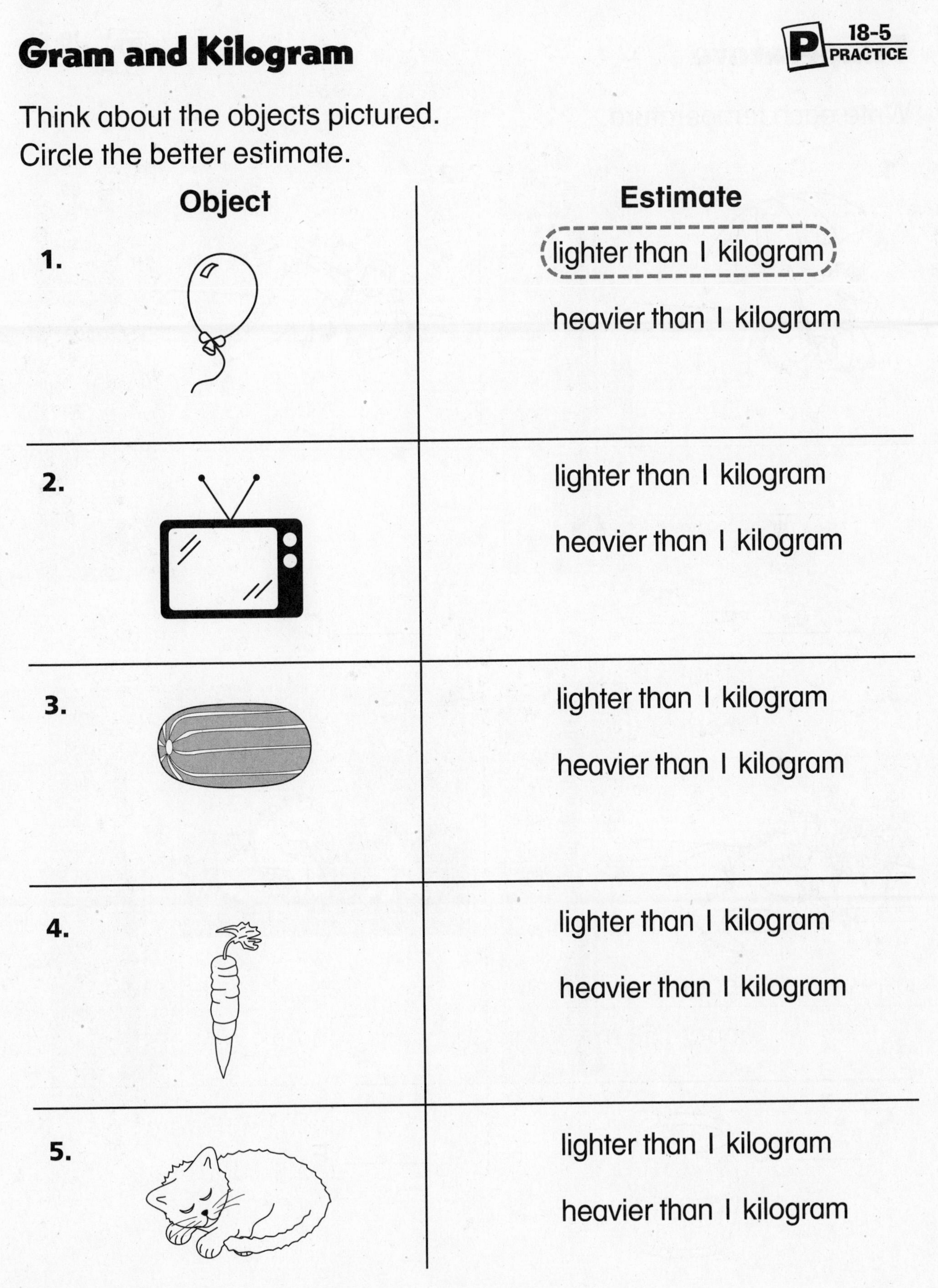

Object	Estimate
1.	lighter than 1 kilogram heavier than 1 kilogram
2.	lighter than 1 kilogram heavier than 1 kilogram
3.	lighter than 1 kilogram heavier than 1 kilogram
4.	lighter than 1 kilogram heavier than 1 kilogram
5.	lighter than 1 kilogram heavier than 1 kilogram

Name ______________________________

Temperature

Write each temperature.

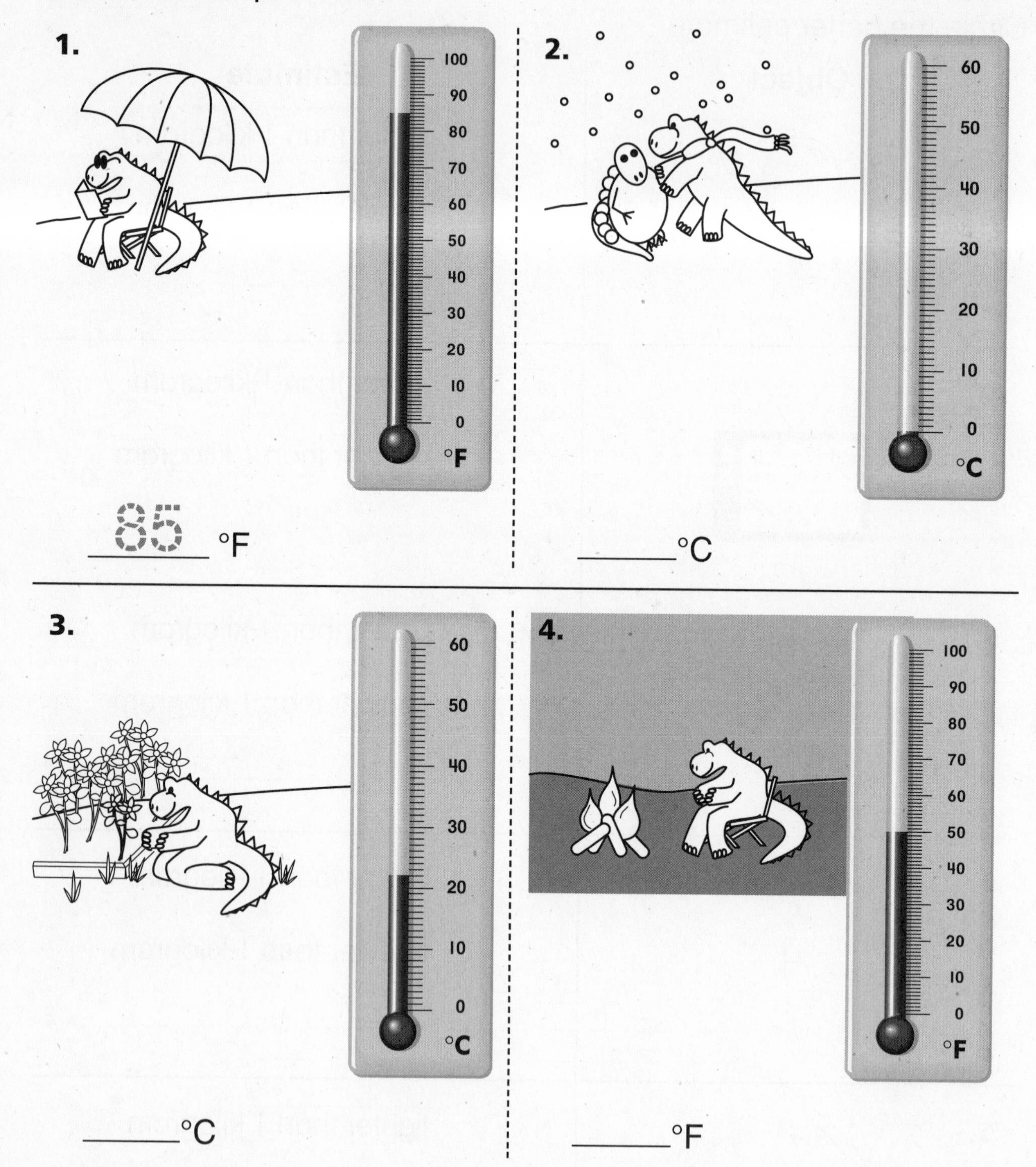

Name ______________________________

18-7 PRACTICE

Problem Solving: Strategy

Use Logical Reasoning

Circle the tool you would use to measure.

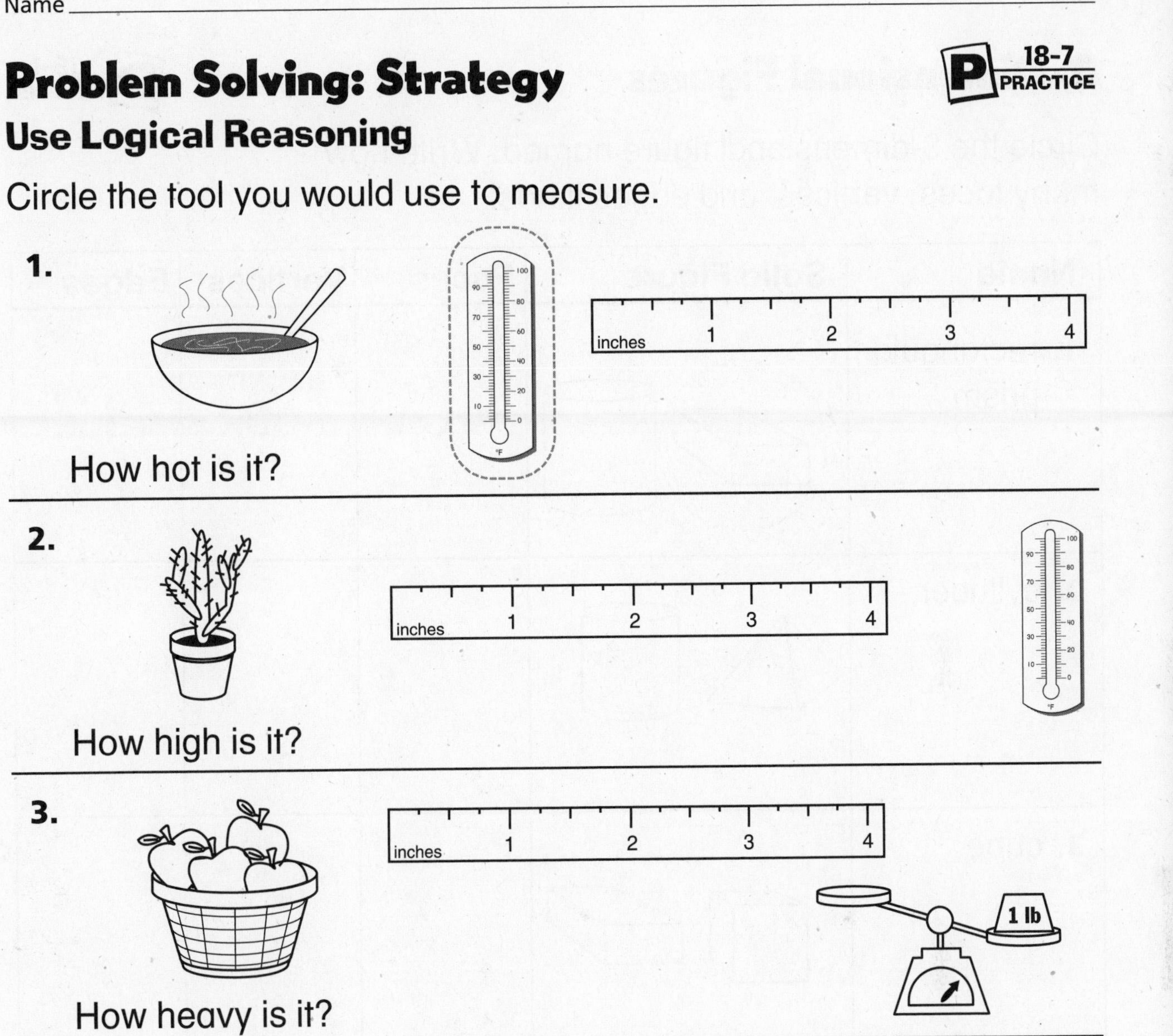

1. How hot is it?

2. How high is it?

3. How heavy is it?

4.

How much water does it hold?

5.

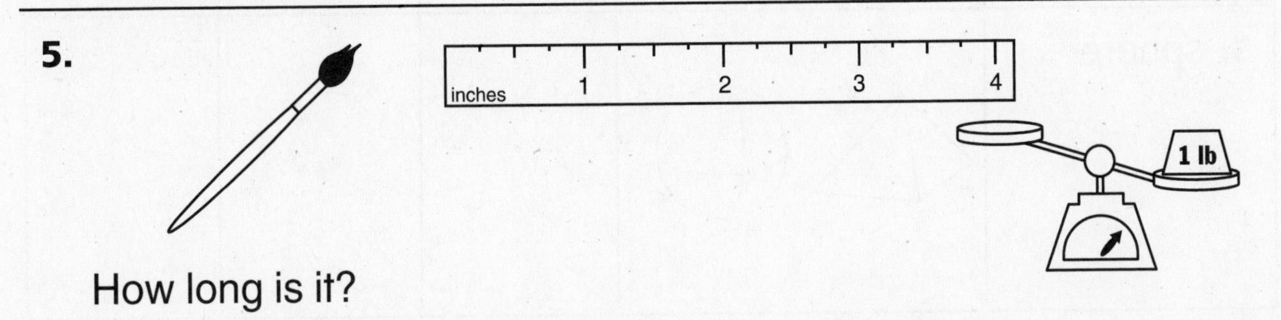

How long is it?

Name ____________________

3-Dimensional Figures

Circle the 3-dimensional figure named. Write how many faces, vertices, and edges it has.

Name	Solid Figure	Faces	Vertices	Edges
1. rectangular prism		6	8	12
2. cylinder				
3. cube				
4. pyramid				
5. sphere				

Use with Grade 2, Chapter 19, Lesson 1, pages 353–354.

Name ____________________

2-Dimensional Shapes

P 19-2 PRACTICE

Color the shape named.
Tell how many sides and angles each has.

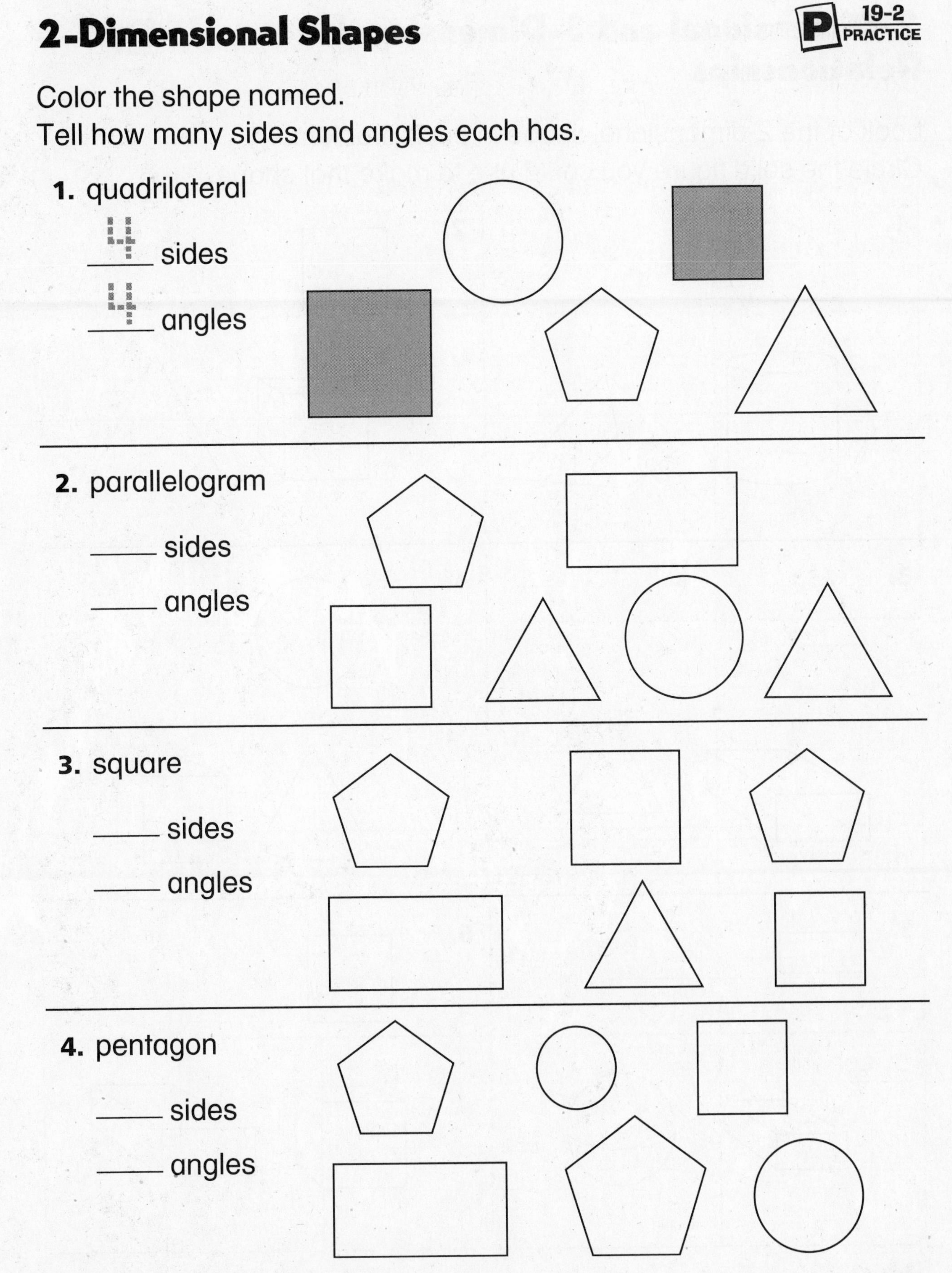

1. quadrilateral

4 sides

4 angles

2. parallelogram

_____ sides

_____ angles

3. square

_____ sides

_____ angles

4. pentagon

_____ sides

_____ angles

Name ____________________

2-Dimensional and 3-Dimensional Relationships

Look at the 2-dimensional shape in each problem.
Circle the solid figure you could use to make that shape.

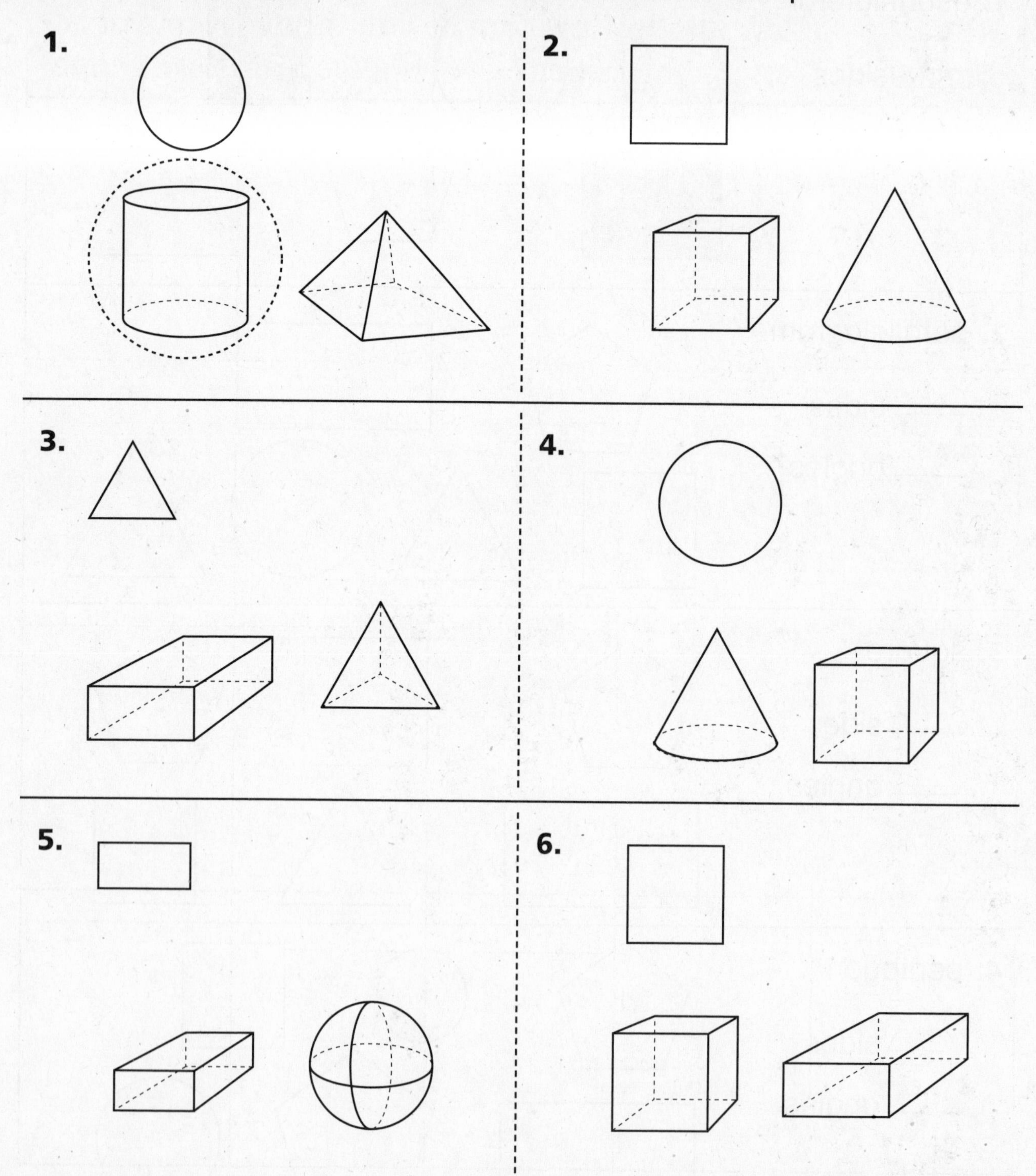

Use with Grade 2, Chapter 19, Lesson 3, pages 357–358.

Name ______________________________

Combine Shapes

Use the pattern blocks to make new shapes.
Complete the chart.

Pattern Blocks	New Shape	How many sides?	How many angles?	Name of New Shape
1.		6	6	hexagon
2.		______	______	
3.		______	______	
4.		______	______	

Name ____________________

Shape Patterns

Use pattern blocks to show the pattern.
Then circle the shapes that repeat.

1.

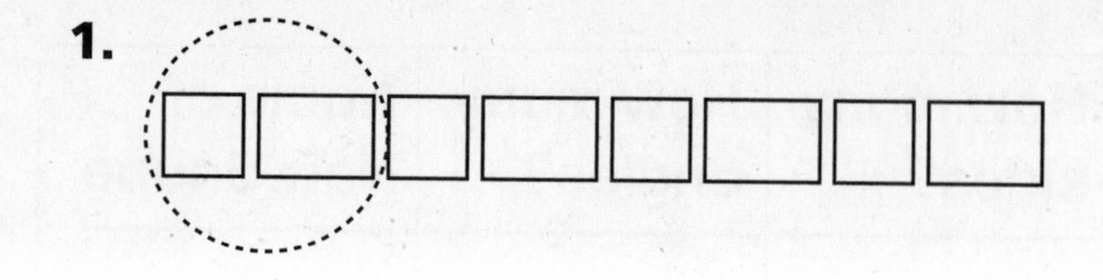

2.

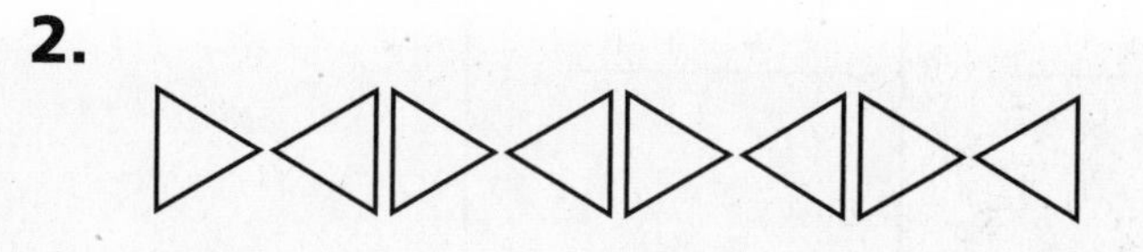

3.

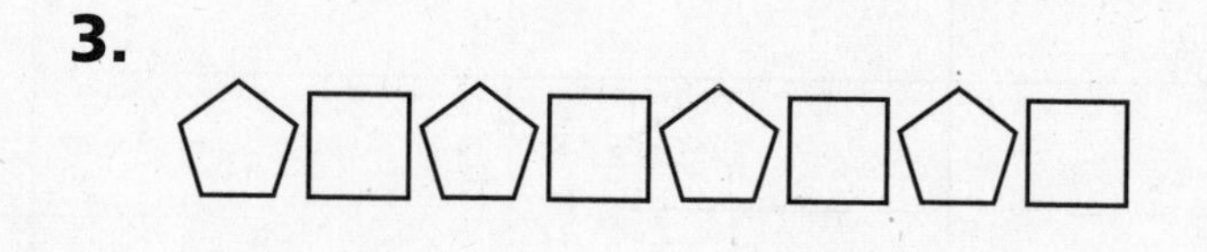

4.

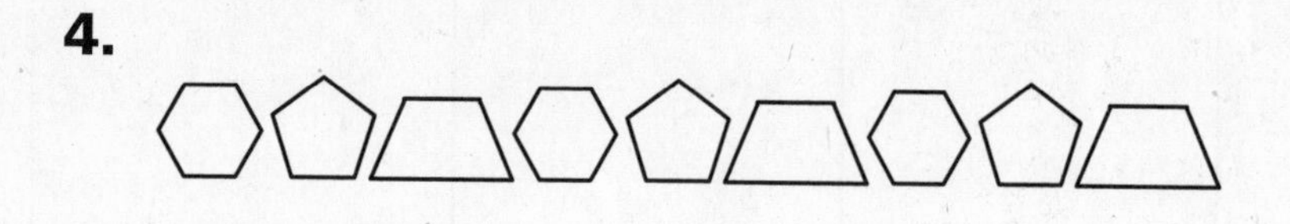

5.

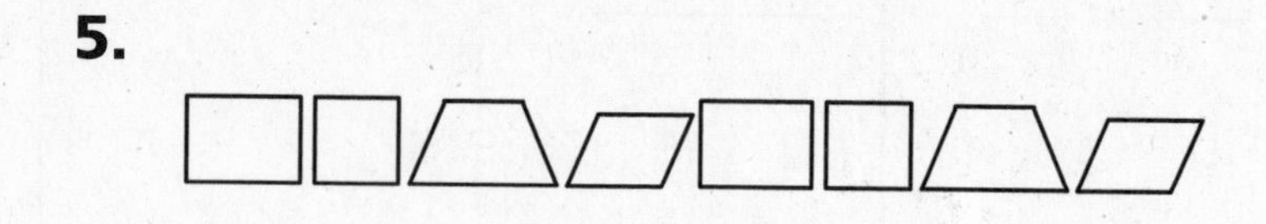

Problem Solving

Solve.

6. Complete the pattern. Draw the missing shape.

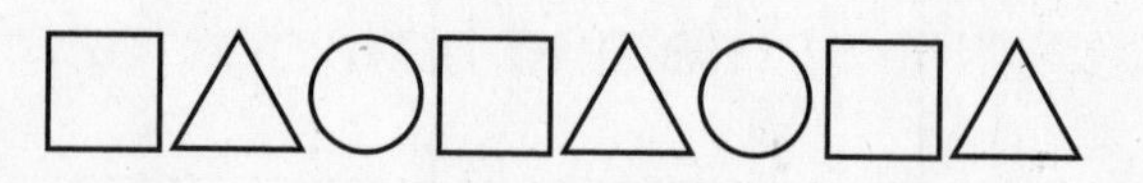

Name ____________________

Problem Solving Skill: Reading for Math

Use Illustrations

The students in Mrs. Park's class are designing a model of their classroom. They use 2-dimensional figures for the chalkboard and posters. They use 3-dimensional figures for the computers, tables, and chalk.

Answer the questions.

1. Which 2-dimensional figures are used for the posters?

____

2. Which 3-dimensional figures are used?

____

3. How are the shapes of the posters the same and how are they different? ____________________

____

Name ____________________

Congruence

20-1 PRACTICE

Color the two congruent shapes in each row.

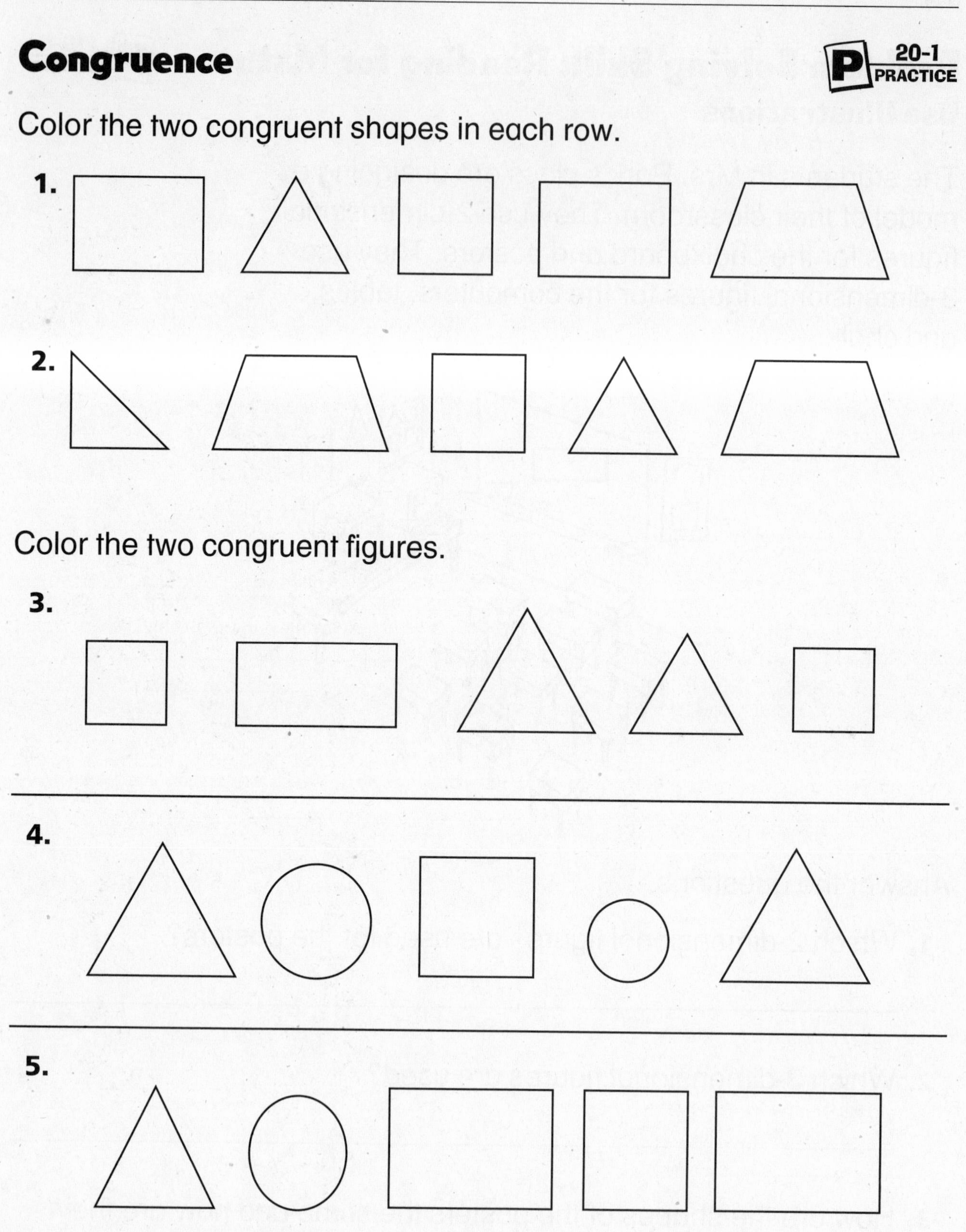

Name ____________________

Symmetry

Draw a matching part for each shape.

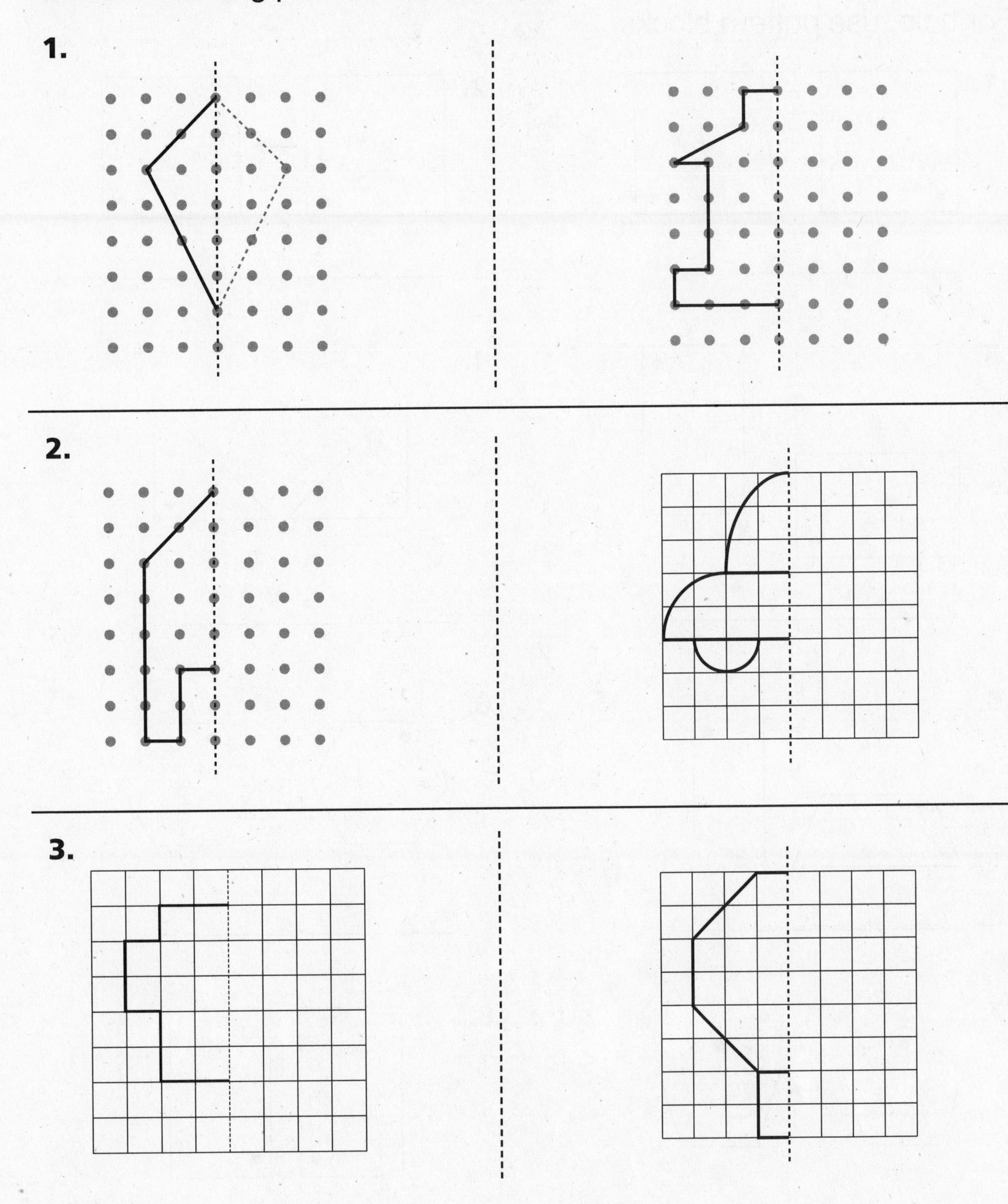

Name ____

Slides, Flips, and Turns

Write the word that names the move.
For help, use pattern blocks.

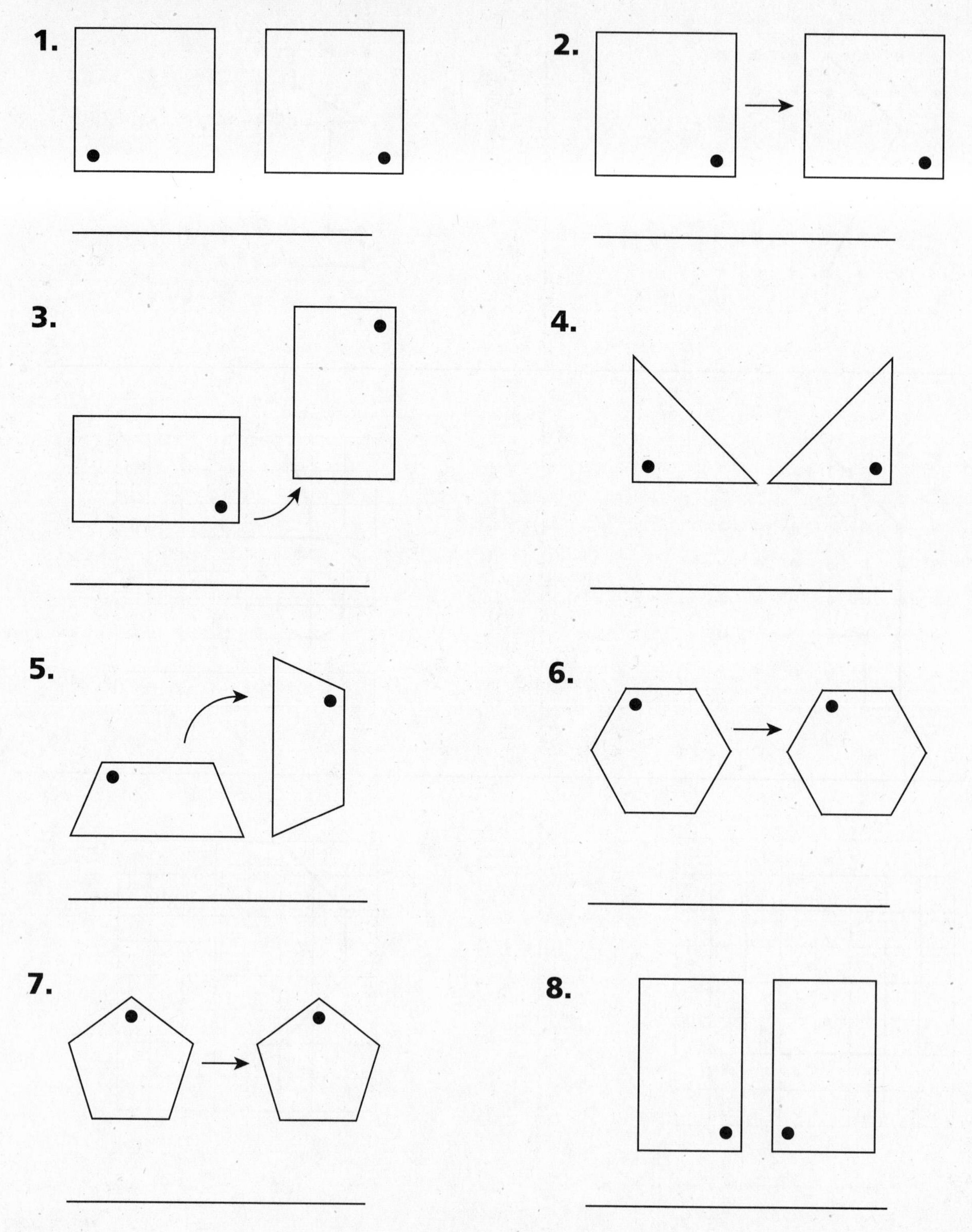

Use with Grade 2, Chapter 20, Lesson 3, pages 377–378.

Name ______________________________

Perimeter

Find the perimeter of each shape. Use an inch ruler.

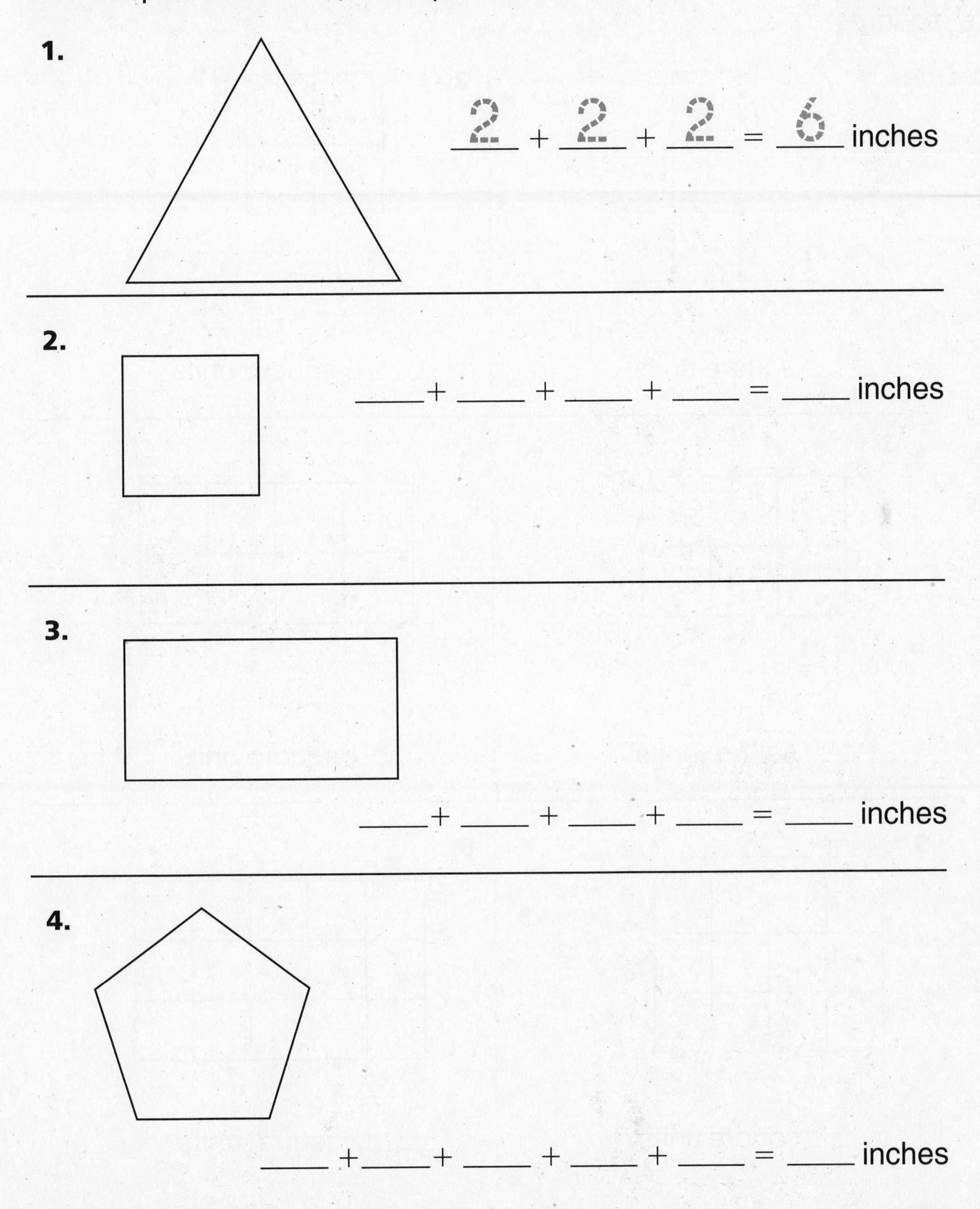

1. 2 + 2 + 2 = 6 inches

2. _____ + _____ + _____ + _____ = _____ inches

3. _____ + _____ + _____ + _____ = _____ inches

4. _____ + _____ + _____ + _____ + _____ = _____ inches

Name ______________________________

Area

Find the area of each shape. Use pattern block squares.

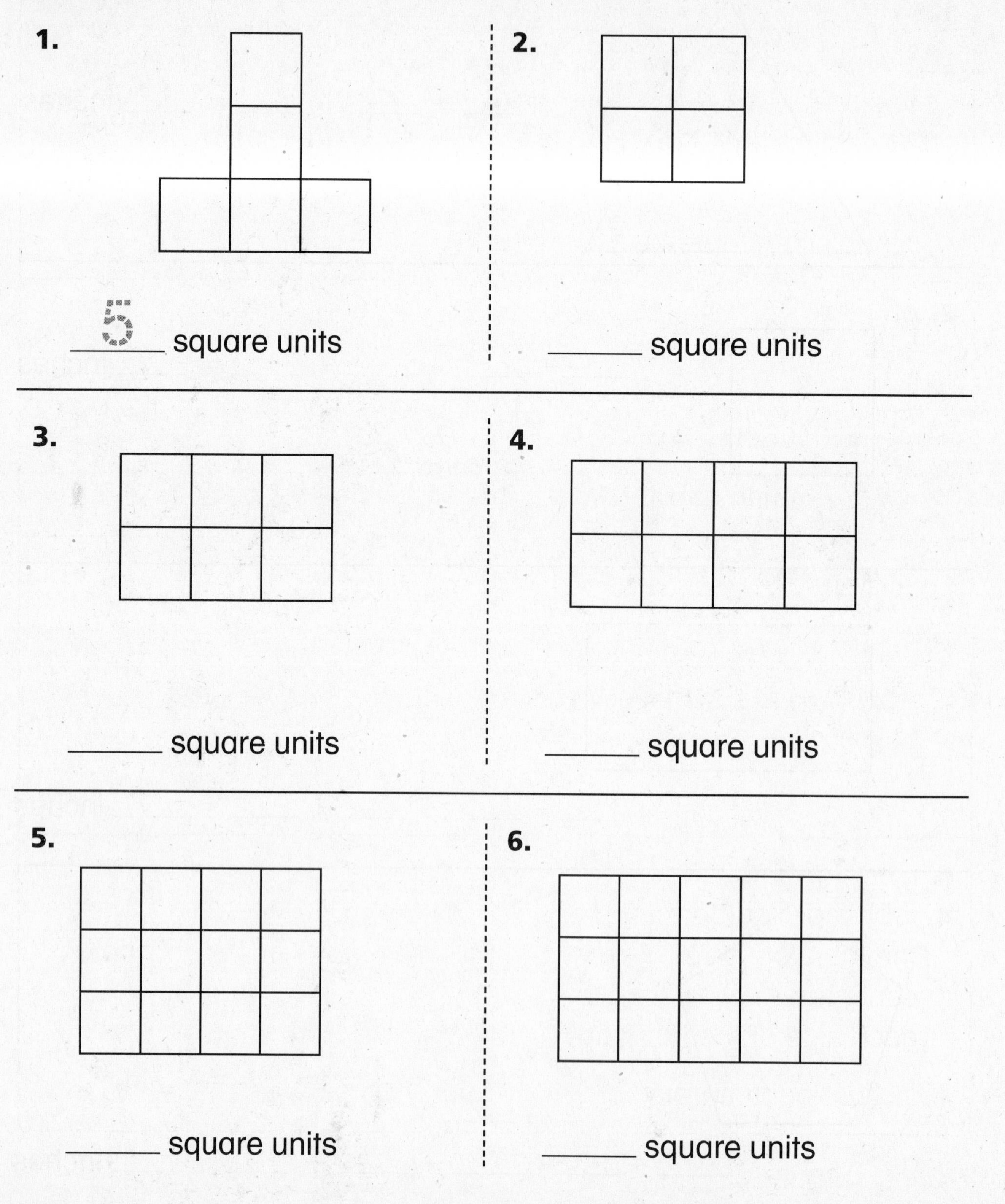

1. 5 square units

2. ______ square units

3. ______ square units

4. ______ square units

5. ______ square units

6. ______ square units

Name ______________________________

Problem Solving: Strategy
Guess and Check

	Draw or write to explain.
1. Lisa draws a square house. The perimeter of the house is 24 centimeters. How long is each side of the house? ______ centimeters	
2. Brian draws a square field. The perimeter of the field is 20 centimeters. How long is each side of the field? ______ centimeters	
3. Keyshawn draws a square building. The perimeter of the building is 28 centimeters. How long is each side of the building? ______ centimeters	
4. Ling draws a square tabletop. The perimeter of the tabletop is 12 centimeters. How long is each side of the tabletop? ______ centimeters	

Name ______________________________

Hundreds

3 groups of one hundred

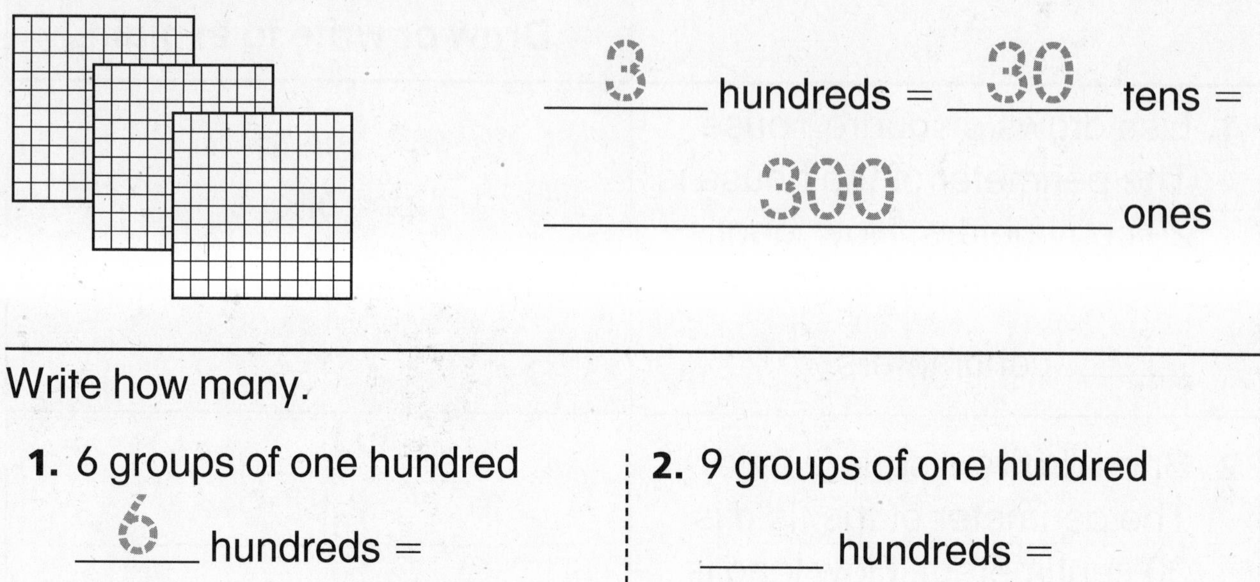

3 hundreds = 30 tens = 300 ones

Write how many.

1. 6 groups of one hundred
   6 hundreds = 60 tens = 600 ones

2. 9 groups of one hundred
   ______ hundreds = ______ tens = ______ ones

3. 4 groups of one hundred
   ______ hundreds = ______ tens = ______ ones

4. 2 groups of one hundred
   ______ hundreds = ______ tens = ______ ones

5. 7 groups of one hundred
   ______ hundreds = ______ tens = ______ ones

6. 1 group of one hundred
   ______ hundred = ______ tens = ______ ones

7. 5 groups of one hundred
   ______ hundreds = ______ tens = ______ ones

8. 8 groups of hundreds
   ______ hundreds = ______ tens = ______ ones

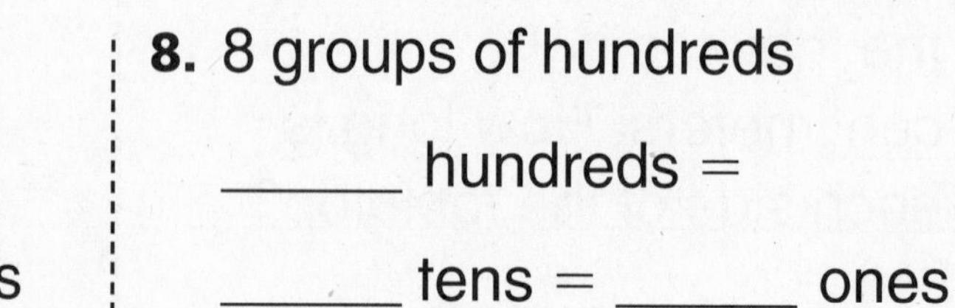

Name ______________________________

Hundreds, Tens, and Ones

You can use hundreds, tens, and ones to show 409.

I must remember to use 0 when there are no tens or ones.

4 hundreds 0 tens 9 ones

Hundreds	Tens	Ones
4	0	9

Write how many hundreds, tens, and ones.

1. 736

____ hundreds ____ tens ____ ones

Hundreds	Tens	Ones

2. 263

____ hundreds ____ tens ____ ones

Hundreds	Tens	Ones

3. 518

____ hundreds ____ ten ____ ones

Hundreds	Tens	Ones

4. 185

____ hundred ____ tens ____ ones

Hundreds	Tens	Ones

5. 360

____ hundreds ____ tens ____ ones

Hundreds	Tens	Ones

Name ______________________________

Place Value Through Hundreds

Write how many hundreds, tens, and ones.
Then write the number.

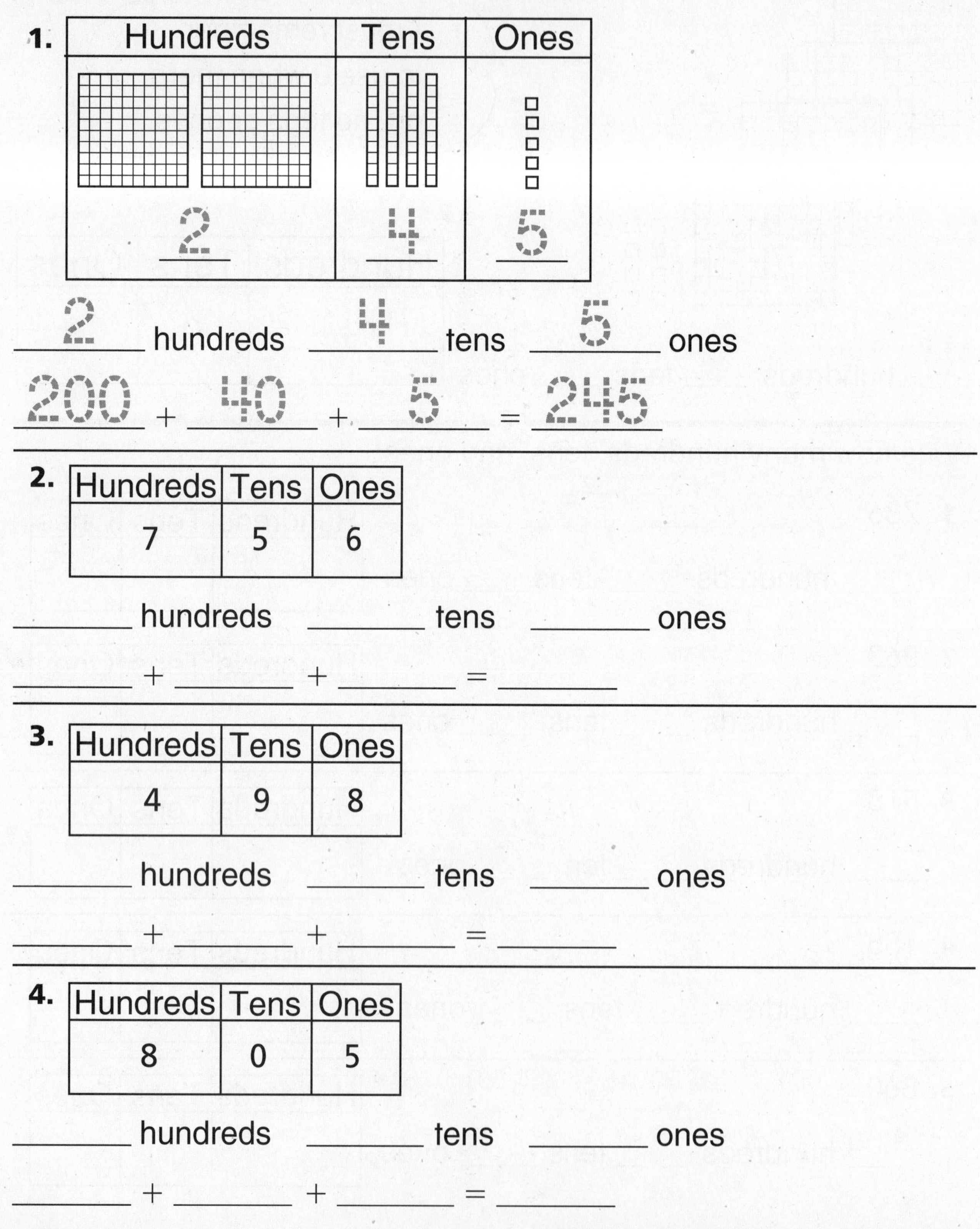

2 hundreds 4 tens 5 ones

200 + 40 + 5 = 245

2.

Hundreds	Tens	Ones
7	5	6

_____ hundreds _____ tens _____ ones

_____ + _____ + _____ = _____

3.

Hundreds	Tens	Ones
4	9	8

_____ hundreds _____ tens _____ ones

_____ + _____ + _____ = _____

4.

Hundreds	Tens	Ones
8	0	5

_____ hundreds _____ tens _____ ones

_____ + _____ + _____ = _____

Name ______________________________

Explore Place Value to Thousands

Write how many thousands, hundreds, tens, and ones. Then write the number.

1.

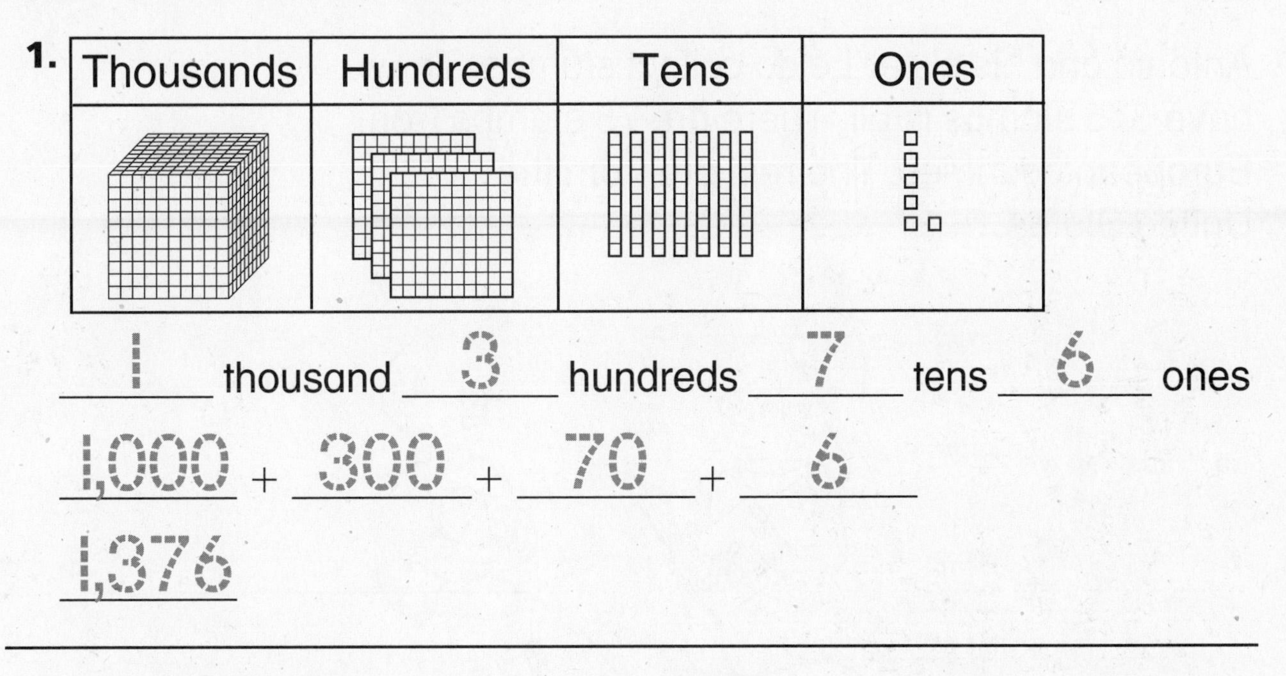

__1__ thousand __3__ hundreds __7__ tens __6__ ones

__1,000__ + __300__ + __70__ + __6__

__1,376__

2.

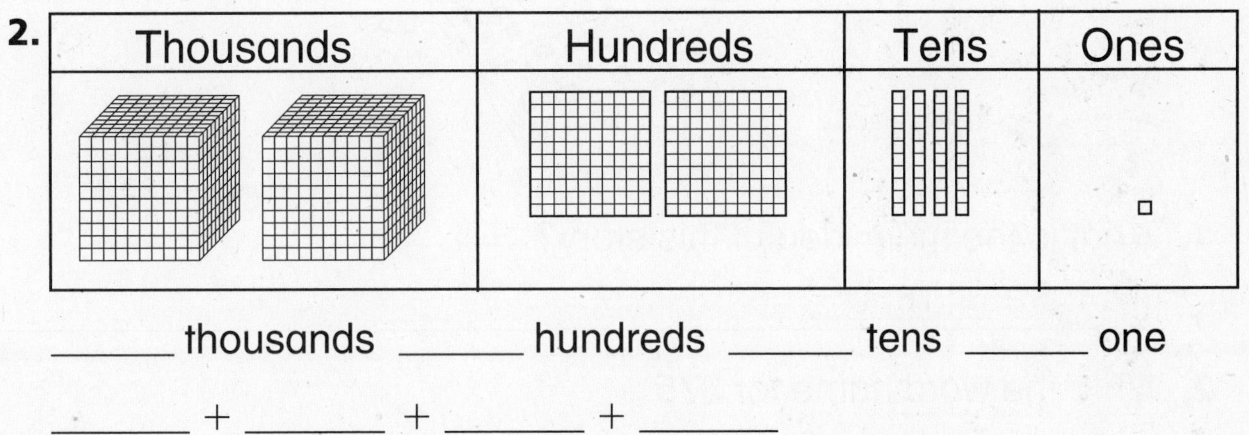

_____ thousands _____ hundreds _____ tens _____ one

_____ + _____ + _____ + _____

_____

3. Use the number line to help you write the missing number in the pattern.

0 1,000 2,000 3,000 4,000 5,000 6,000 7,000 8,000 9,000 10,000

3,000, 4,000, _____, _____, 7,000, _____, 9,000

Name ___

Problem Solving Skill: Reading for Math

Find the Main Idea

Read the story.

Antonio and his sister Leida collect stamps. They have 375 stamps in all. There are 75 stamps from European countries. The rest are from the United States.

1. What is the main idea of this story?

___

2. Write the word name for 375.

___

3. Write the number 375 in expanded form.

___ + ___ + ___

4. Antonio has 250 stamps and Leida has 125 stamps.

Who has more stamps? ___

Name ______________________________

Compare Numbers • Algebra

Compare. Write >, <, or =.

1.	415 (<) 451	623 ( ) 678	730 ( ) 830
2.	375 ( ) 375	549 ( ) 560	248 ( ) 239
3.	109 ( ) 111	382 ( ) 379	445 ( ) 545
4.	272 ( ) 275	818 ( ) 816	357 ( ) 357
5.	643 ( ) 637	256 ( ) 261	429 ( ) 421
6.	317 ( ) 371	588 ( ) 598	761 ( ) 769
7.	285 ( ) 287	638 ( ) 632	954 ( ) 957
8.	275 ( ) 375	717 ( ) 717	539 ( ) 542
9.	827 ( ) 789	690 ( ) 711	431 ( ) 438
10.	555 ( ) 525	684 ( ) 648	698 ( ) 698

Name

Order Numbers on a Number Line

Write the number that is just before and just after.

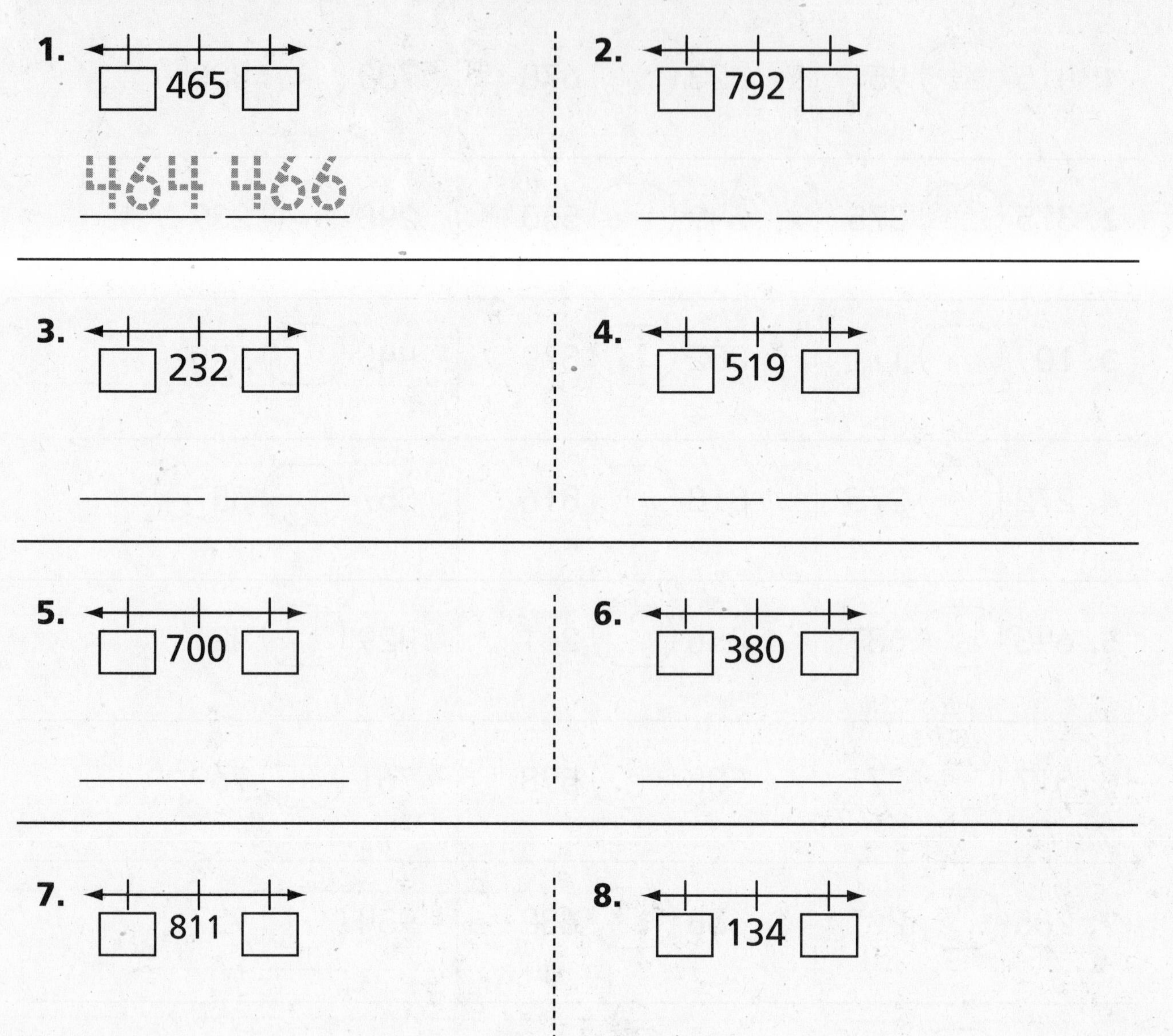

Use with Grade 2, Chapter 22, Lesson 2, pages 417–418.

Name ___

Order Numbers

22-3 PRACTICE

Order the numbers from least to greatest.

1. 274, 248, 312, 291   248, 274, 291, 312

2. 682, 628, 631, 619   ___, ___, ___, ___

3. 485, 554, 444, 452   ___, ___, ___, ___

4. 712, 638, 824, 722   ___, ___, ___, ___

Order the numbers from greatest to least.

5. 387, 235, 412, 370   412, 387, 370, 235

6. 919, 901, 991, 109   ___, ___, ___, ___

7. 832, 328, 283, 823   ___, ___, ___, ___

8. 164, 192, 187, 148   ___, ___, ___, ___

Name ______________________________

Number Patterns • Algebra

22-4 PRACTICE

Write the missing numbers.
Then circle the pattern.

Count by:

1. 715, 725, 735, 745, 755 hundreds tens ones

2. 673, ______, 675, ______, 677 hundreds tens ones

3. 491, ______, 691, ______, 891 hundreds tens ones

4. ______, 839, ______, 841, 842 hundreds tens ones

5. ______, 229, ______, 429, 529 hundreds tens ones

6. 548, 648, ______, ______, 948 hundreds tens ones

7. ______, 395, 495, 595, ______ hundreds tens ones

8. 579, 589, 599, ______, ______ hundreds tens ones

Name ______________________________

Count Forward, Count Backward

P 22-5 PRACTICE

Count forward or backward by ones.

Count forward by ones. Count backward by ones.

1. 439, 440, 441 394, 393, 392

2. 924, ______, ______ 182, ______, ______

3. 599, ______, ______ 711, ______, ______

Count forward or backward by tens.

Count forward by tens. Count backward by tens.

4. 824, 834, 844 267, 257, 247

5. 782, ______, ______ 182, ______, ______

6. 599, ______, ______ 711, ______, ______

Count forward or backward by hundreds.

Count forward by hundreds. Count backward by hundreds.

7. 188, 288, 388 750, 650, 550

8. 507, ______, ______ 427, ______, ______

9. 646, ______, ______ 999, ______, ______

Name ______________________________

Problem Solving: Strategy

Make a Table

Solve.
Complete the tables.

1. Mindy puts a nickel in her bank each day.
How much can she save in 5 days?

Days	1				
Money	5¢	¢	¢	¢	¢

25 ¢

2. Sue bought 5 boxes of tissues. Each box has 100 sheets.
How many tissues did she buy?

Boxes					
Tissues					

_______ tissues

3. Tim buys 5 packs of sports cards. There are 10 cards in each pack. How many cards does he buy?

Packs					
Cards					

_______ cards

4. The theater has 5 sections of seats. There are 50 seats in each section. How many seats are there?

Sections					
Seats					

_______ seats

Use with Grade 2, Chapter 22, Lesson 6, pages 425–426.

Name ______________________________

P 23-1 PRACTICE

Count on by hundreds to add.

1. $300 + 100 = 400$ $200 + 400$ $700 + 100$ $200 + 300$ $100 + 400$

2. $600 + 100$ $300 + 200$ $600 + 200$ $800 + 100$ $400 + 500$

3. $500 + 200$ $400 + 200$ $500 + 300$ $200 + 200$ $100 + 700$

4. $300 + 300$ $400 + 300$ $100 + 100$ $200 + 200$ $100 + 800$

Problem Solving

Solve.

Show Your Work

5. There are 400 students in the second grade. There are 400 students in the third grade. How many students are there in all?

_______ students

Name ____________________

Regroup Ones

23-2 PRACTICE

Use hundreds, tens, and ones to add.

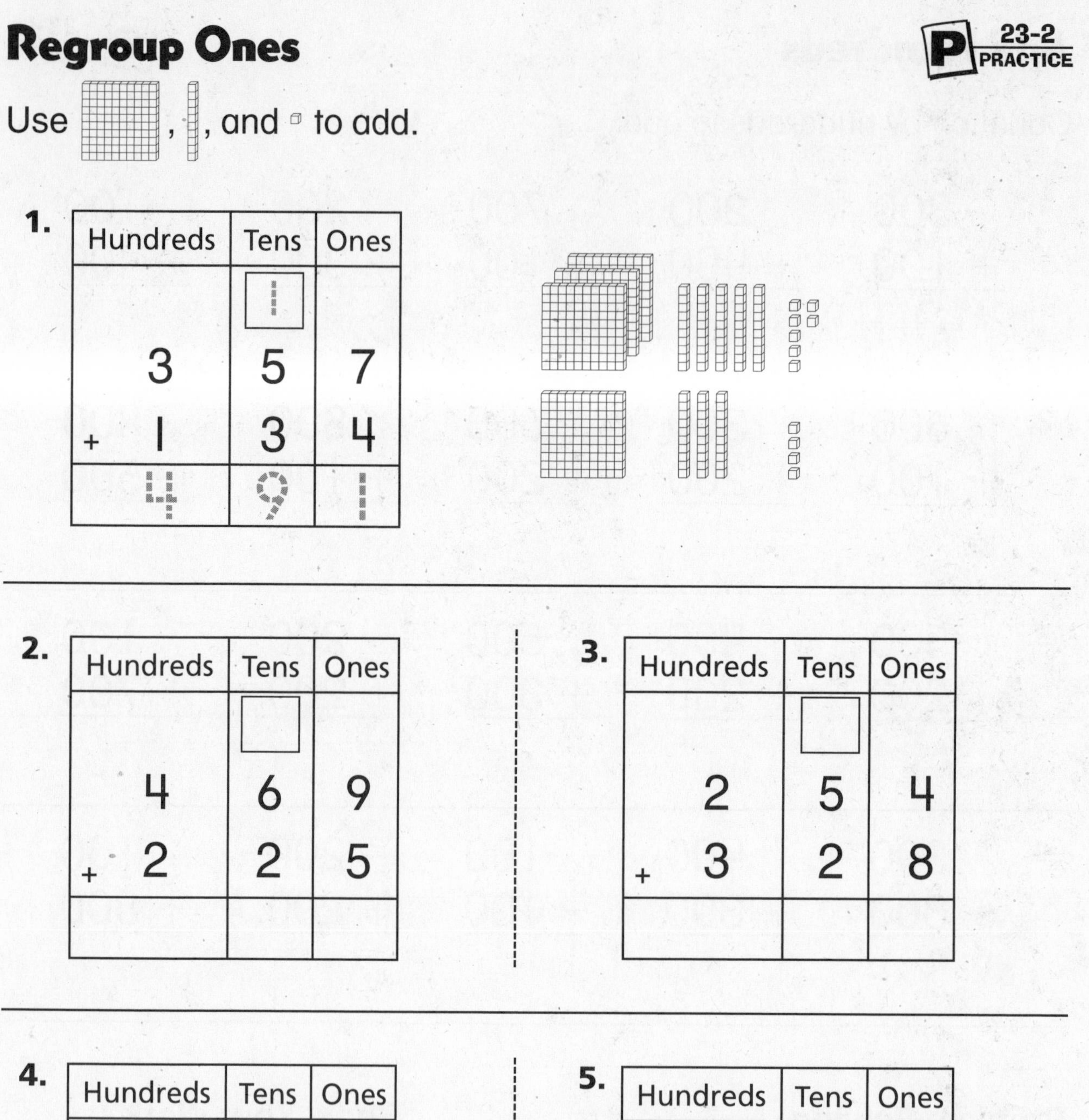

1.

Hundreds	Tens	Ones
	1	
3	5	7
+ 1	3	4
4	9	1

2.

Hundreds	Tens	Ones
	☐	
4	6	9
+ 2	2	5

3.

Hundreds	Tens	Ones
	☐	
2	5	4
+ 3	2	8

4.

Hundreds	Tens	Ones
	☐	
2	5	8
+ 7	2	9

5.

Hundreds	Tens	Ones
5	3	2
+ 2	5	5

Name ______________________________

Regroup Tens

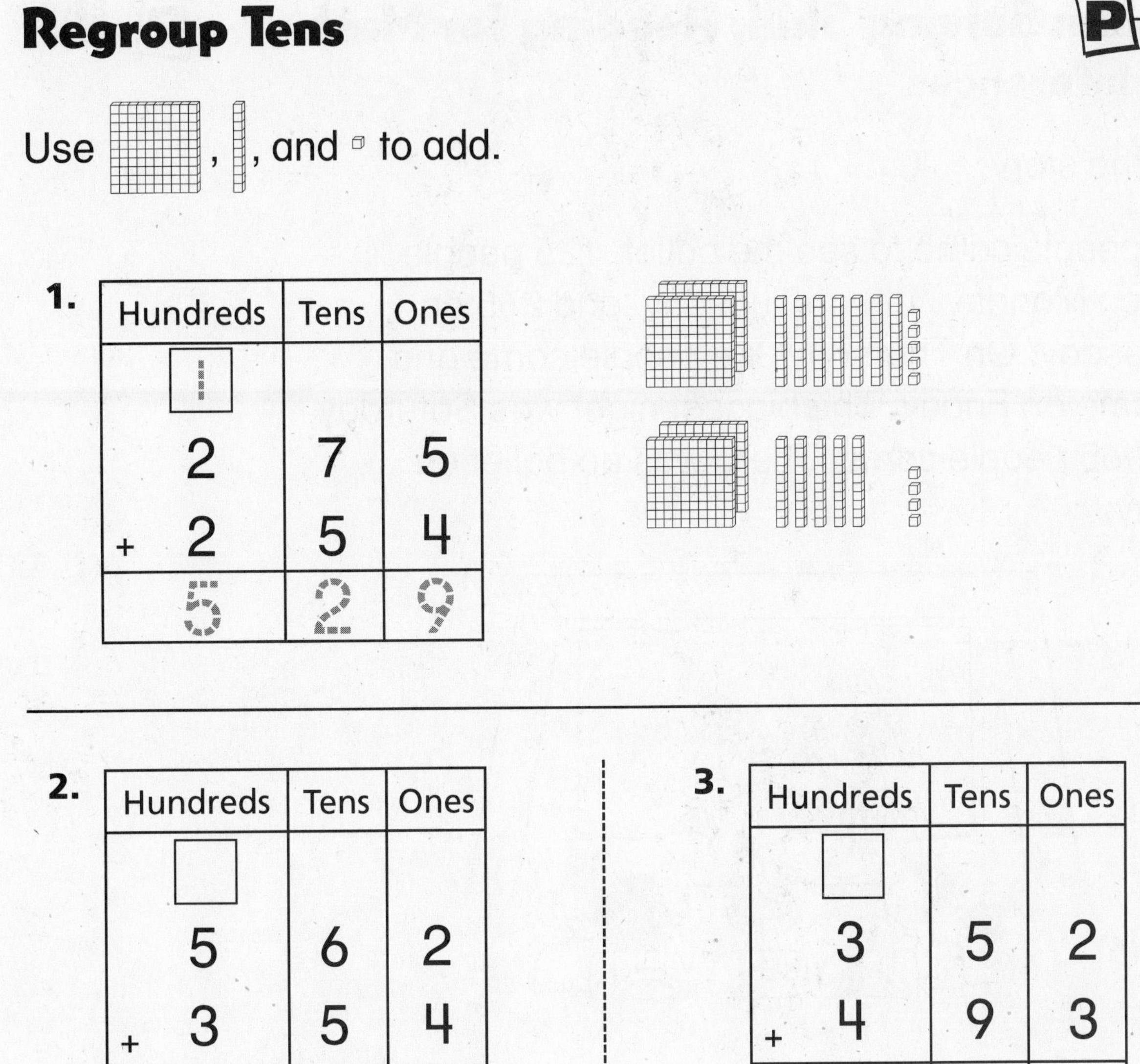

Use hundreds, tens, and ones to add.

1.

Hundreds	Tens	Ones
1		
2	7	5
+ 2	5	4
5	2	9

2.

Hundreds	Tens	Ones
☐		
5	6	2
+ 3	5	4

3.

Hundreds	Tens	Ones
☐		
3	5	2
+ 4	9	3

4.

Hundreds	Tens	Ones
☐		
4	5	3
+ 1	7	5

5.

Hundreds	Tens	Ones
4	3	4
+ 3	1	5

Name ____________________

Problem Solving Skill: Reading for Math

Make Inferences

Read the story.

Many people came to see the ballet. 125 people came on Monday, 140 on Tuesday, and 265 on Wednesday. On Thursday 326 people came and 450 came on Friday. The biggest night was Saturday, when 465 people came. There was no ballet on Sunday.

1. How many people saw the ballet on Monday and Tuesday?

   ____

2. How many people saw the ballet on Wednesday and Thursday? ____________________

3. How many people saw the ballet on Friday and Saturday? ____________________

4. Why do you think more people saw the ballet on Saturday than on Thursday?

   ____

   ____

Use with Grade 2, Chapter 23, Lesson 4, pages 441–442.

Name ______________________________

Subtract Hundreds

24-1 PRACTICE

Subtract.

1.	$\begin{array}{r} 300 \\ -\ 100 \\ \hline 200 \end{array}$	$\begin{array}{r} 800 \\ -\ 300 \\ \hline \end{array}$	$\begin{array}{r} 700 \\ -\ 100 \\ \hline \end{array}$	$\begin{array}{r} 600 \\ -\ 300 \\ \hline \end{array}$	$\begin{array}{r} 600 \\ -\ 200 \\ \hline \end{array}$
2.	$\begin{array}{r} 400 \\ -\ 100 \\ \hline \end{array}$	$\begin{array}{r} 500 \\ -\ 100 \\ \hline \end{array}$	$\begin{array}{r} 600 \\ -\ 500 \\ \hline \end{array}$	$\begin{array}{r} 800 \\ -\ 100 \\ \hline \end{array}$	$\begin{array}{r} 500 \\ -\ 300 \\ \hline \end{array}$
3.	$\begin{array}{r} 500 \\ -\ 200 \\ \hline \end{array}$	$\begin{array}{r} 900 \\ -\ 200 \\ \hline \end{array}$	$\begin{array}{r} 600 \\ -\ 400 \\ \hline \end{array}$	$\begin{array}{r} 700 \\ -\ 400 \\ \hline \end{array}$	$\begin{array}{r} 800 \\ -\ 500 \\ \hline \end{array}$
4.	$\begin{array}{r} 400 \\ -\ 300 \\ \hline \end{array}$	$\begin{array}{r} 800 \\ -\ 600 \\ \hline \end{array}$	$\begin{array}{r} 700 \\ -\ 300 \\ \hline \end{array}$	$\begin{array}{r} 900 \\ -\ 300 \\ \hline \end{array}$	$\begin{array}{r} 900 \\ -\ 100 \\ \hline \end{array}$

Problem Solving

Solve.

Show Your Work

5. 900 children are in the park.
700 adults are in the park.
How many more children are there than adults?

______ more children

Name ___

Regroup Tens as Ones

Use [hundred flat], [ten rod], and [one cube] to subtract.

hundreds	tens	ones
	6	12
2	~~7~~	~~2~~
− 1	3	6
1	3	6

1.

hundreds	tens	ones
	☐	☐
7	6	3
− 3	2	5

2.

hundreds	tens	ones
	☐	☐
6	5	7
− 4	2	9

3.

hundreds	tens	ones
	☐	☐
4	8	3
− 1	2	8

4.

hundreds	tens	ones
	☐	☐
8	6	1
− 5	4	5

5.

hundreds	tens	ones
	☐	☐
5	4	6
− 4	2	9

6.

hundreds	tens	ones
9	5	7
− 7	2	5

Name ____________________

Regroup Hundreds as Tens

24-3 PRACTICE

Use , , and to subtract.

hundreds	tens	ones
2	12	
3	2	8
− 2	7	7
	5	1

1.

hundreds	tens	ones
☐	☐	
5	6	7
− 2	9	5

2.

hundreds	tens	ones
☐	☐	
9	1	2
− 5	6	2

3.

hundreds	tens	ones
☐	☐	
7	2	7
− 3	8	2

4.

hundreds	tens	ones
☐	☐	
8	3	8
− 4	4	5

5.

hundreds	tens	ones
☐	☐	
3	3	9
− 1	6	8

6.

hundreds	tens	ones
8	9	7
− 6	4	6

Name ________________________________

Estimate, Add, and Subtract Money Amounts

P 24-4 PRACTICE

Add or subtract. Estimate to see if your answer is reasonable.

1.

		nearest dollar			
	\$3.29	nearest dollar	\$3.00	\$5.82	\$6.00
+	4.43	nearest dollar	+ 4.00	− 3.67	− 4.00
	\$7.72	about	\$7.00	\$2.15	\$2.00

2.

\$5.39 − 2.73	− ____	\$2.74 + 5.21	+ ____

3.

\$3.91 + 1.73	+ ____	\$7.14 − 2.71	− ____

Problem Solving

Solve.

4. Timo has \$6.85. He wants to buy a book for \$3.58. He estimates he will still have enough money left to buy a magazine. Is Timo right? Explain your answer.

____

____

____

____

Name ________________________________

Problem Solving: Strategy

Work Backward

24-5 PRACTICE

Solve.

Draw or write to explain.

1. Paula is reading a book with 346 pages. She has 152 pages left to read. How many pages has she read already?

   194 pages

   346
   − 152
   194

2. Greta has 875 marbles. She has 692 left after she gives some to Brenda. How many marbles did Brenda get?

   _______ marbles

3. Jorge bought a book for $4.35. Now he has $3.29. How much money did Jorge have to begin with?

   ____

4. There are 658 books in Tom's collection. Tom donates some to the library. Now he has 485 books. How many books did Tom donate to the library?

   _______ books

Name ____________________

Unit Fractions

P 25-1 PRACTICE

Write the fraction for the shaded part.

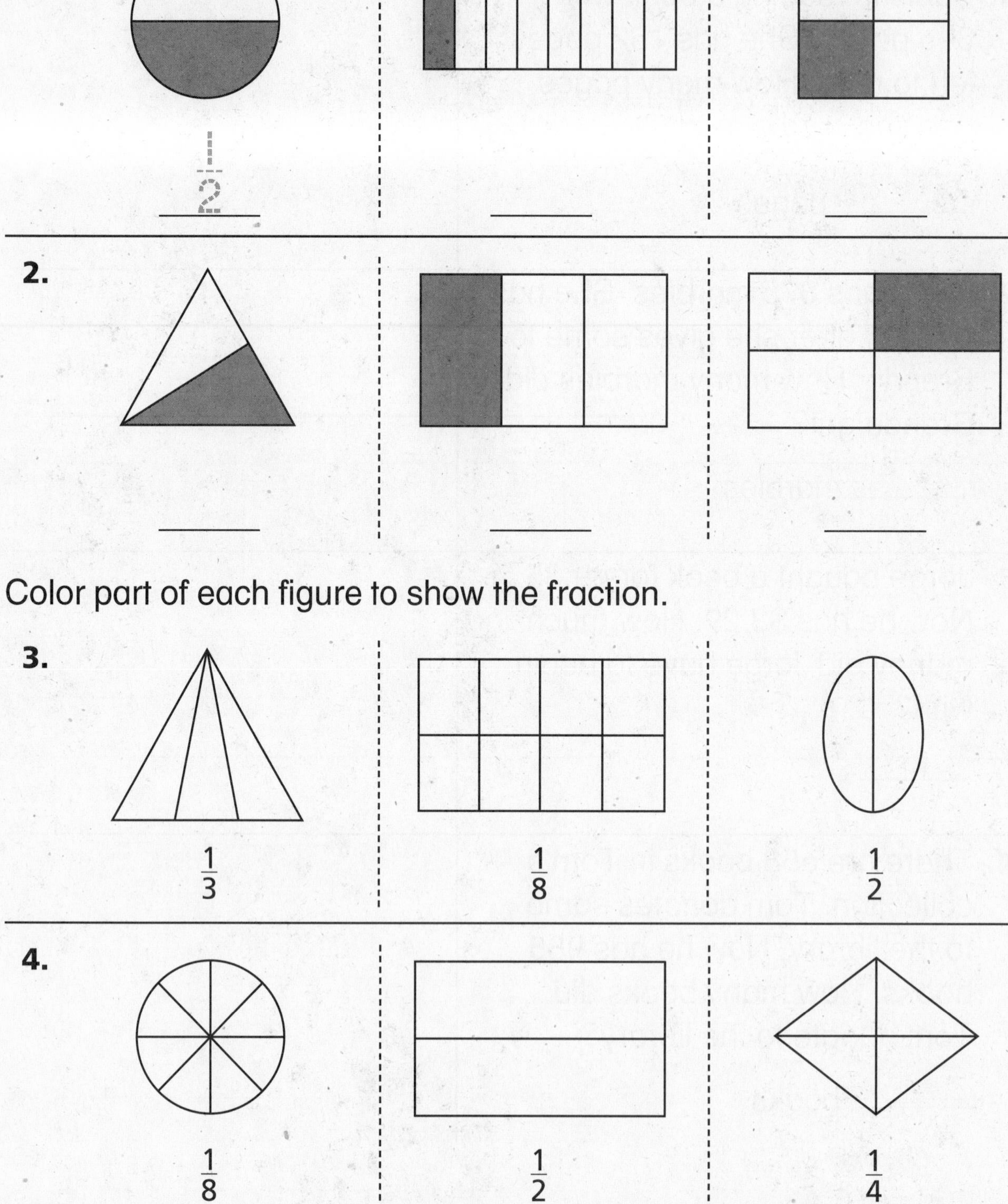

Use with Grade 2, Chapter 25, Lesson 1, pages 473–474.

Name ____________________

Fractions Equal to 1

Count the parts in each whole.
Color the parts using the same color crayon.
Then write the fraction for the whole.

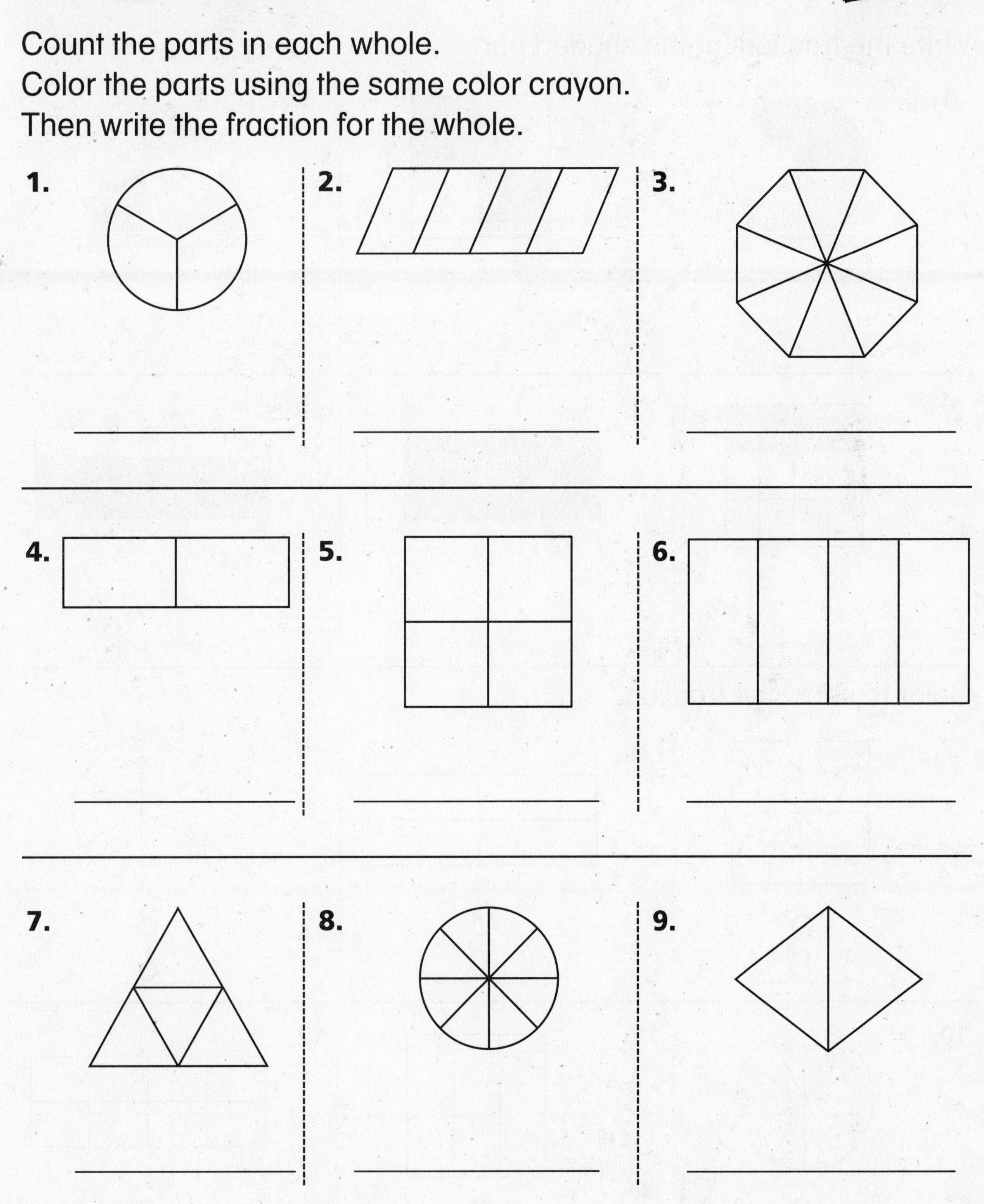

Name ____________________

Other Fractions

25-3 PRACTICE

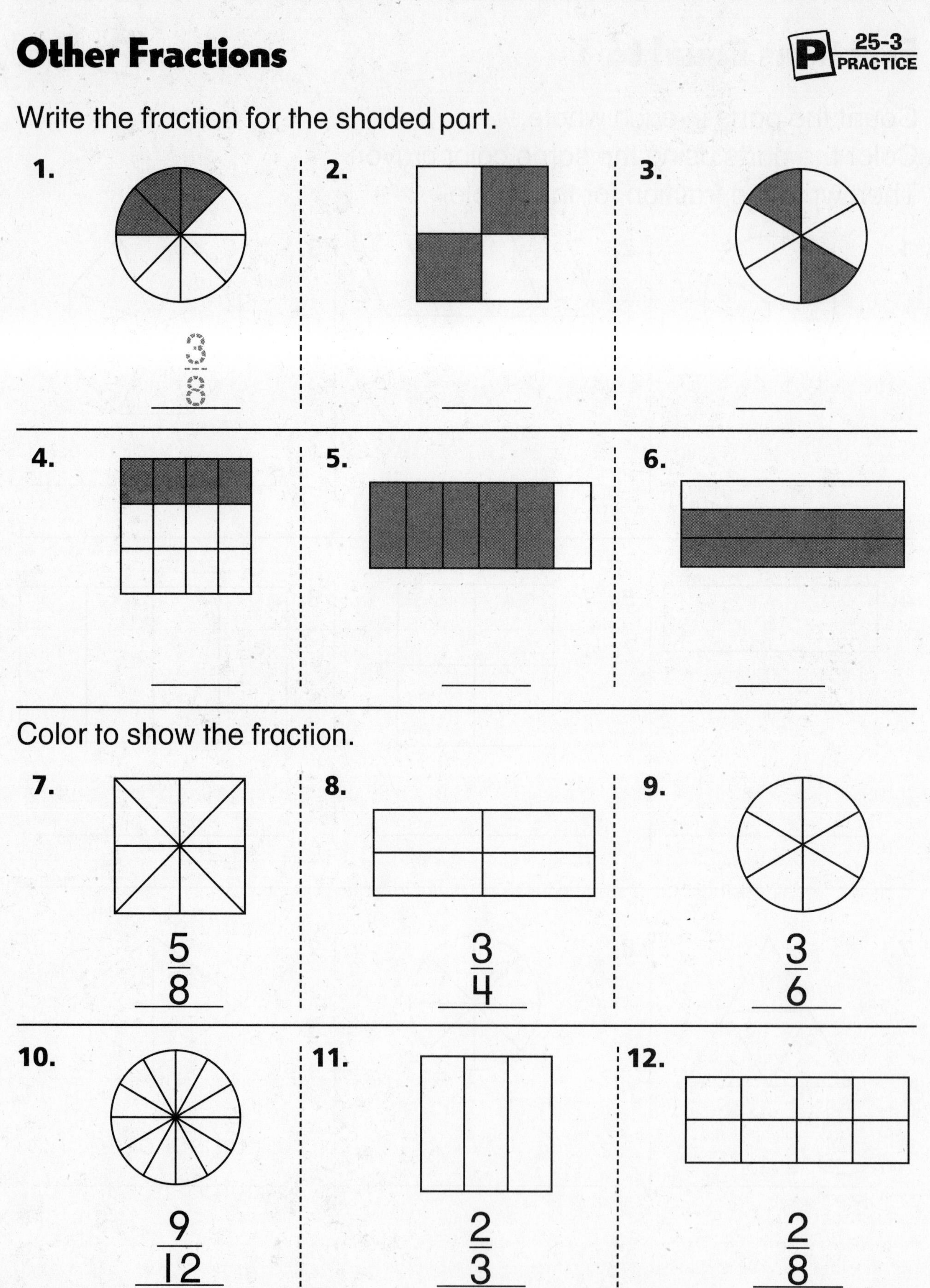

Write the fraction for the shaded part.

1. $\frac{3}{8}$

2. ______

3. ______

4. ______

5. ______

6. ______

Color to show the fraction.

7. $\frac{5}{8}$

8. $\frac{3}{4}$

9. $\frac{3}{6}$

10. $\frac{9}{12}$

11. $\frac{2}{3}$

12. $\frac{2}{8}$

Name ______________________________

Unit Fractions of a Group

P 25-4 PRACTICE

Write the fraction for the shaded part.

1. $\frac{5}{6}$

2. ______

3. ______

Look at the picture. Write the fraction.

4. What fraction of the animals are starfish?

2 → total number of starfish
4 → total number of animals

5. What fraction of the animals are goldfish?

______ → total number of goldfish
→ total number of animals

6. What fraction of the animals are dolphins?

______ → total number of dolphins
→ total number of animals

Use with Grade 2, Chapter 25, Lesson 4, pages 479–480.

Name ____________________

Other Fractions of a Group

25-5 PRACTICE

Color to show the fraction of the group.

1. $\frac{5}{6}$ of the fish are green.

2. $\frac{3}{4}$ of the fish are pink.

3. $\frac{3}{3}$ of the fish are blue.

4. $\frac{3}{8}$ of the fish are red.

5. $\frac{1}{2}$ of the fish are yellow.

6. $\frac{1}{6}$ of the fish are orange.

Use the picture. Write the fraction.

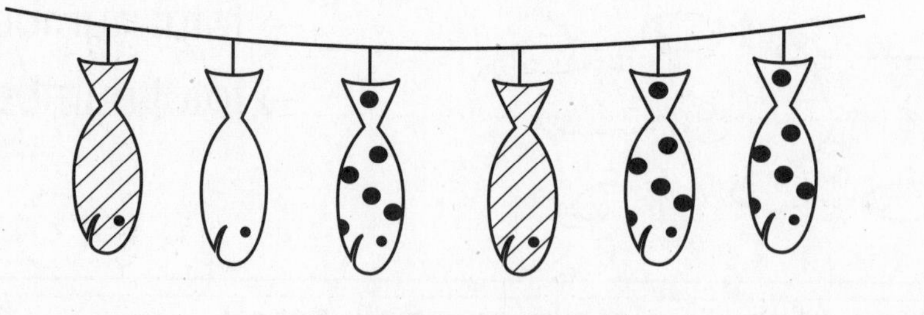

7. ____ of the fish are striped.

8. ____ of the fish are spotted.

9. ____ of the fish are white.

Name ____________________

Compare Fractions

P 25-6 PRACTICE

Compare the fractions. Then write $<$ or $>$.

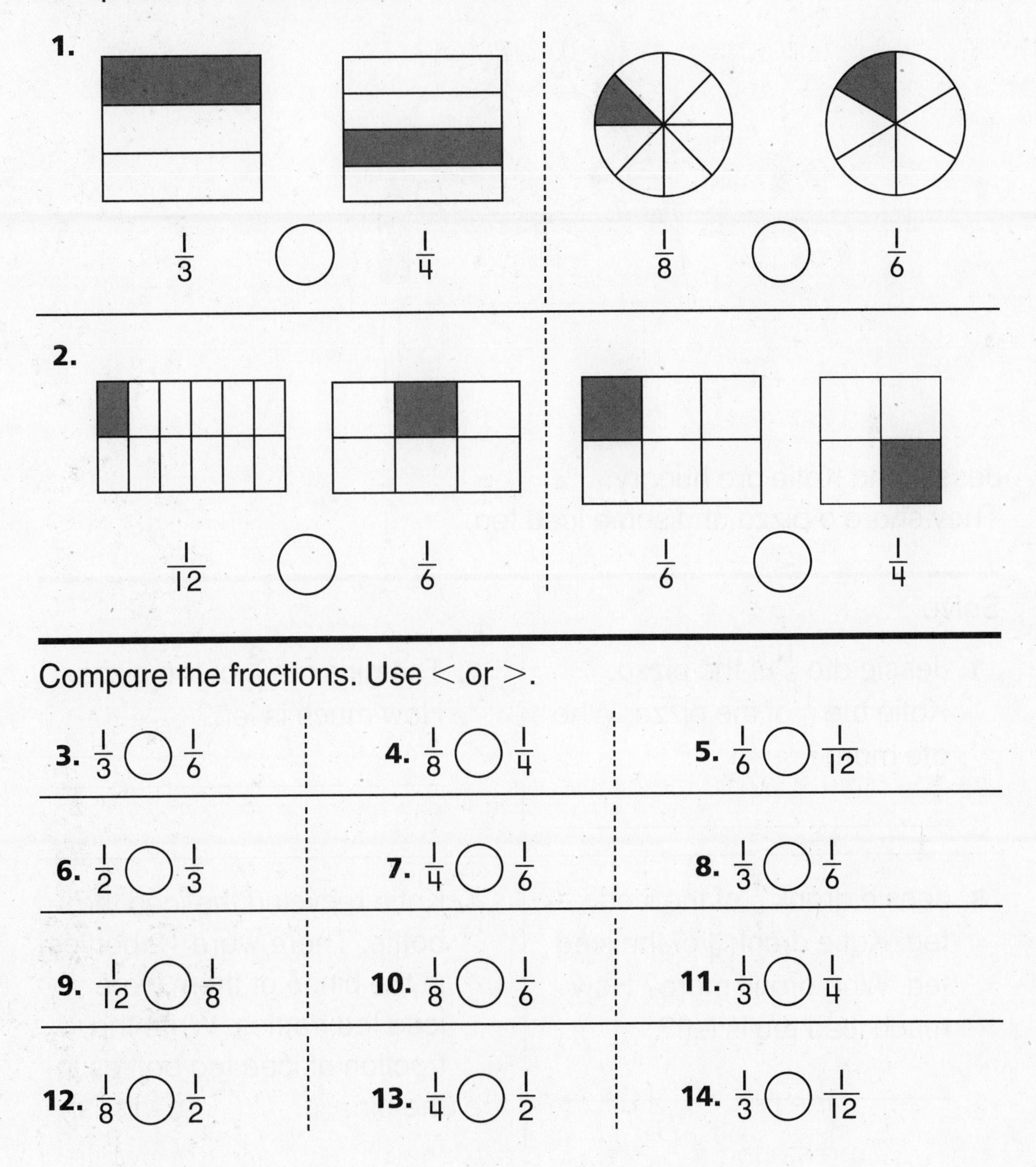

1. $\frac{1}{3}$ ◯ $\frac{1}{4}$ $\frac{1}{8}$ ◯ $\frac{1}{6}$

2. $\frac{1}{12}$ ◯ $\frac{1}{6}$ $\frac{1}{6}$ ◯ $\frac{1}{4}$

Compare the fractions. Use $<$ or $>$.

3. $\frac{1}{3}$ ◯ $\frac{1}{6}$

4. $\frac{1}{8}$ ◯ $\frac{1}{4}$

5. $\frac{1}{6}$ ◯ $\frac{1}{12}$

6. $\frac{1}{2}$ ◯ $\frac{1}{3}$

7. $\frac{1}{4}$ ◯ $\frac{1}{6}$

8. $\frac{1}{3}$ ◯ $\frac{1}{6}$

9. $\frac{1}{12}$ ◯ $\frac{1}{8}$

10. $\frac{1}{8}$ ◯ $\frac{1}{6}$

11. $\frac{1}{3}$ ◯ $\frac{1}{4}$

12. $\frac{1}{8}$ ◯ $\frac{1}{2}$

13. $\frac{1}{4}$ ◯ $\frac{1}{2}$

14. $\frac{1}{3}$ ◯ $\frac{1}{12}$

Name ______________________________

Problem Solving Skill: Reading for Math

Draw Conclusions

Jessie and Katie are hungry.
They share a pizza and some iced tea.

Solve.

1. Jessie ate $\frac{1}{8}$ of the pizza. Katie ate $\frac{1}{4}$ of the pizza. Who ate more?

____

2. The girls ate $\frac{3}{8}$ of the pizza. How much is left?

____

3. Jessie drank $\frac{1}{4}$ of the iced tea. Katie drank $\frac{3}{4}$ of the iced tea. Who drank more? How much iced tea is left?

____

4. Katie recycled the iced tea bottle. There were 12 bottles in the bin. 6 of them were iced tea bottles. Write the fraction of iced tea bottles in the bin.

____

Name ______________________________

Explore Probability

Use the spinners to answer the questions.
Circle the answers.

1. That the spinner will land on stripes is ____.

certain
probable
impossible

2. That the spinner will land on spots is ____.

certain
probable
impossible

3. That the spinner will land on white is ____.

certain
probable
impossible

4. That the spinner will land on white is ____.

certain
probable
impossible

Problem Solving

Solve.

5. What color will you be certain to pick?

____

6. What color is impossible to pick?

____

green
green
green

Name ____________________

More Likely, Equally Likely, or Less Likely

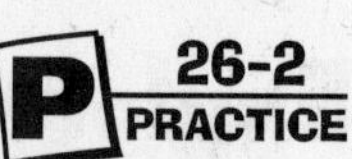

Look at each picture. Suppose the cubes are in the bag.

Without looking, which color are you more likely to pick?

Which color are you less likely to pick?

Color the cube.

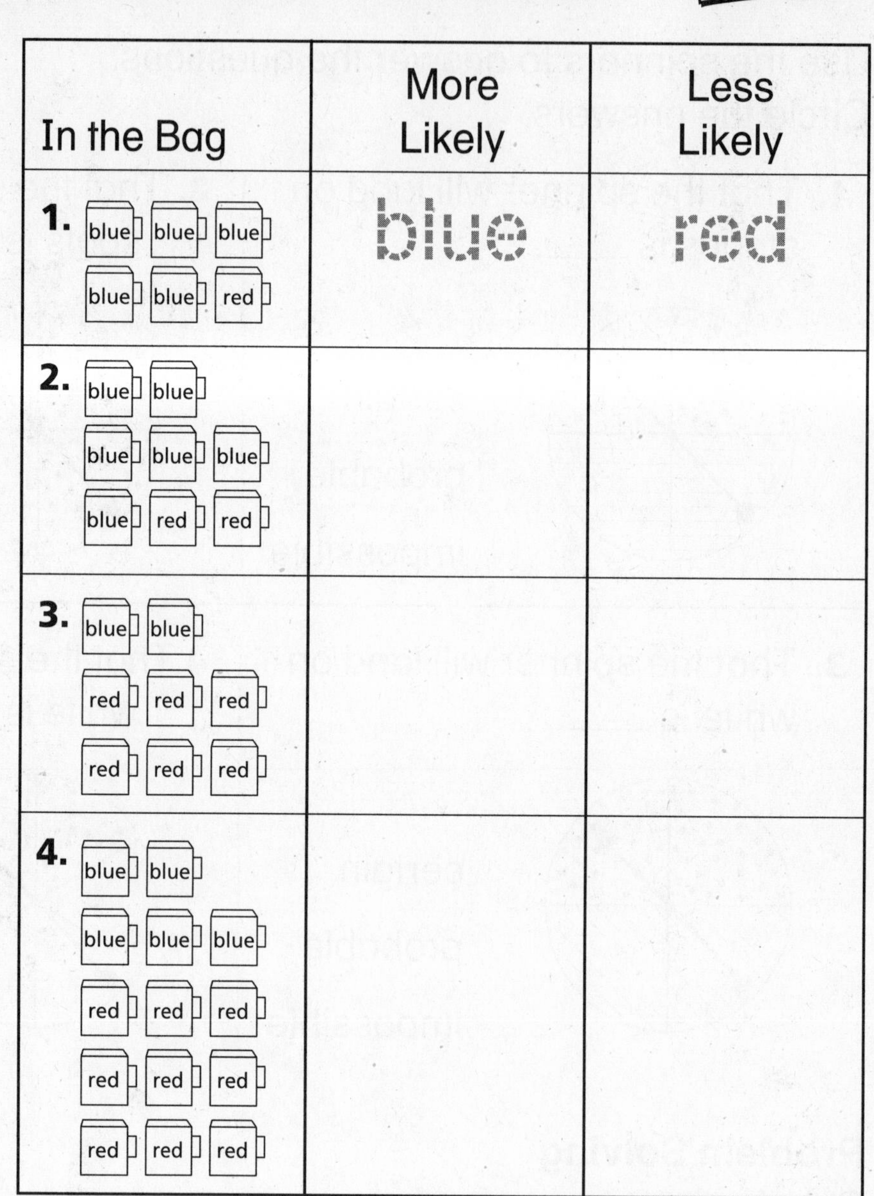

In the Bag	More Likely	Less Likely
1. blue blue blue blue blue red	blue	red
2. blue blue blue blue blue blue red red		
3. blue blue red red red red red red		
4. blue blue blue blue blue red red red red red red red red red		

Problem Solving

Solve.

5. Color the cubes using red and blue so that red or blue are equally likely to be picked.

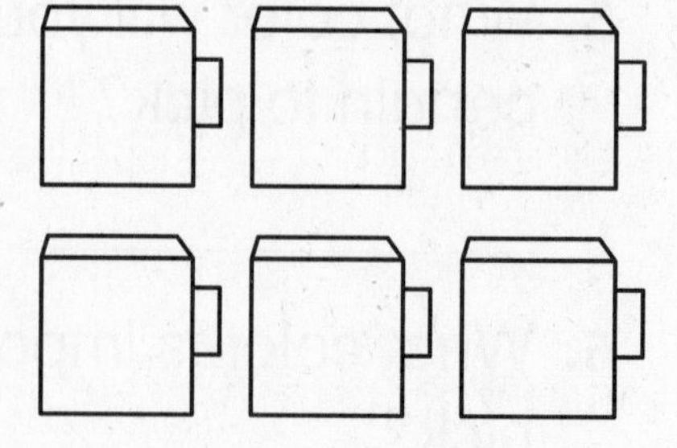

Name ____________________

Make Predictions

Use the spinner to answer the questions.

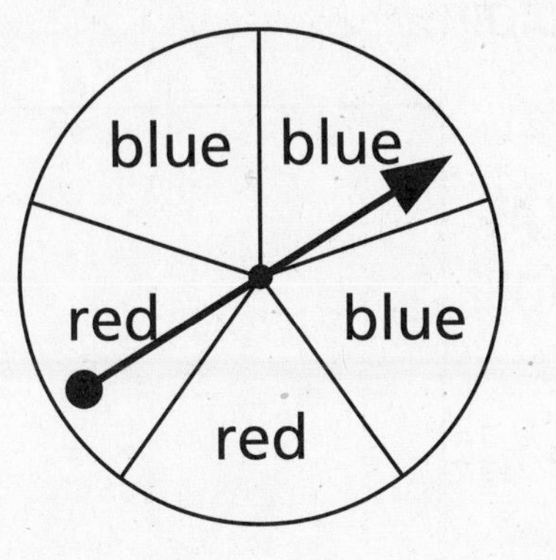

1. What color do you predict you will spin more often? ____________
2. Why do you think so? ____________
3. Predict. If you spin the spinner 10 times, how many times will you get red? ____________

Use red and blue to make your own spinner.

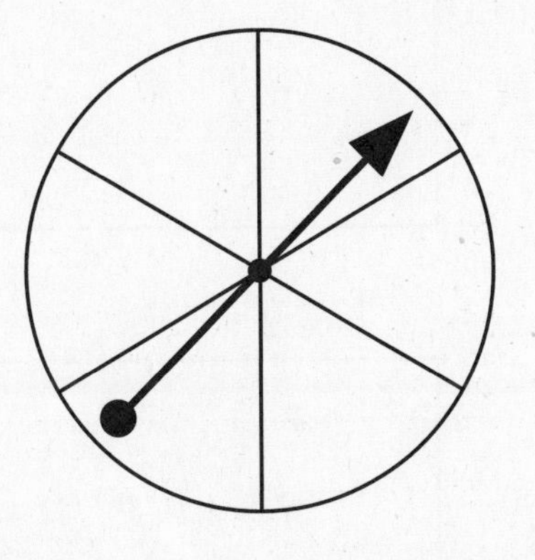

4. What color do you predict you will spin more often? ____________
5. Why do you think so? ____________
6. Predict. If you spin the spinner 10 times, how many times will you get red? ____________

Name ______________________________

Problem Solving: Strategy

Make a List

Complete. Use the space to explain.

1. Kobe uses the numbers 3, 5, and 9 to make different 2-digit numbers. What numbers can she make, using each digit only once in a number?

35, 39, _____, _____,

_____, _____

2. Cathy uses the numbers 1, 4, and 6 to make different 2-digit numbers. What numbers can she make, using each digit only once in a number?

_____, _____, _____, _____,

_____, _____

3. Debra made different 2-digit numbers with 2, 3, and 7, using each digit only once in a number. What numbers did she make?

_____, _____, _____, _____,

_____, _____

Name ____________________

Range and Mode

27-1 PRACTICE

Use the data. Answer each question.

Tim's team scored this many points in the last 5 games.

2 5 9 4 4

1. What is the range of the numbers? 7

2. What is the mode of the numbers? 4

Annie's team scored this many points in 6 games.

1 6 3 3 9 5

3. What is the range of the numbers? ____________

4. What is the mode of the numbers? ____________

This is the number of children on each team.

22 25 25 21 25

5. What is the range of the numbers? ____________

6. What is the mode of the numbers? ____________

This many bags of popcorn were sold at the last 6 games.

37 21 44 44 32 27

7. What is the range of the numbers? ____________

8. What is the mode of the numbers? ____________

Name ____

Median

Make towers with following number of cubes.

Find the median.

1. 12 cubes, 7 cubes, 9 cubes, 4 cubes, 10 cubes

Median: ___9___ cubes

---

2. 4 cubes, 8 cubes, 8 cubes, 7 cubes, 6 cubes

Median: ________ cubes

---

3. 13 cubes, 11 cubes, 14 cubes, 15 cubes, 10 cubes

Median: ________ cubes

---

4. 9 cubes, 10 cubes, 12 cubes, 8 cubes, 12 cubes

Median: ________ cubes

Problem Solving
Solve.

Show Your Work

5. Lucie and her friends walked the following number of miles in the past 5 days.

2 3 1 1 4

What is the median of the numbers?

____

Name ______________________________

Coordinate Graphs • Algebra

Where is each animal?
Write the numbers.

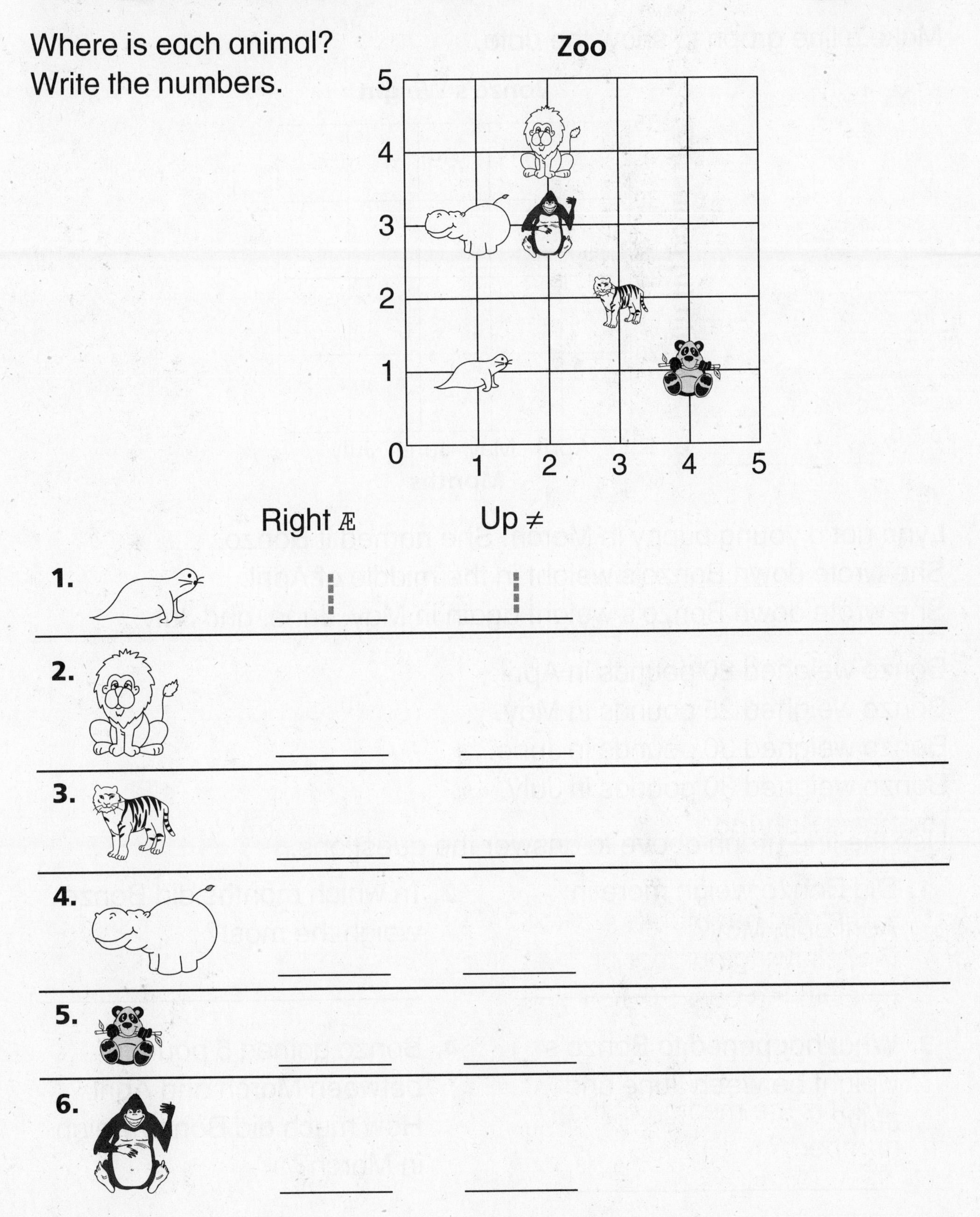

	Right Æ	Up ≠
1.	1	1
2.	______	______
3.	______	______
4.	______	______
5.	______	______
6.	______	______

Name ____

Line Graphs

Make a line graph to show the data.

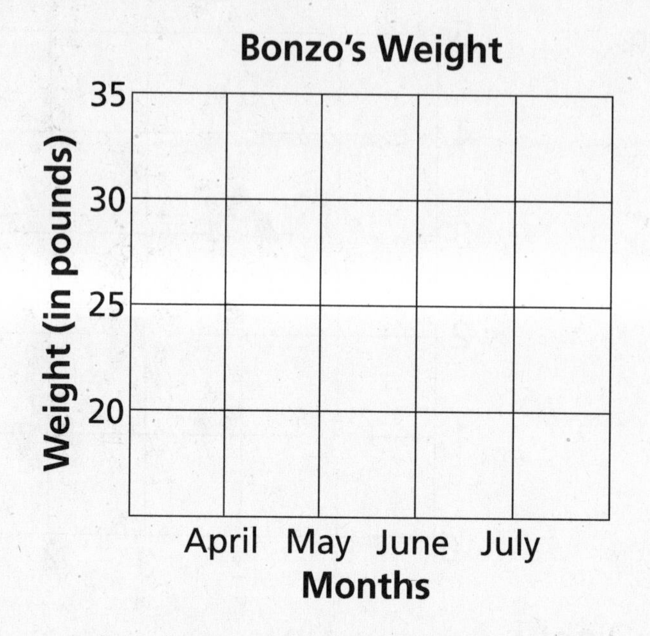

Lynn got a young puppy in March. She named it Bonzo.
She wrote down Bonzo's weight in the middle of April.
She wrote down Bonzo's weight again in May, June, and July.

Bonzo weighed 20 pounds in April.
Bonzo weighed 25 pounds in May.
Bonzo weighed 30 pounds in June.
Bonzo weighed 30 pounds in July.

Use the line graph above to answer the questions.

1. Did Bonzo weigh more in April or in May?

____

2. In which months did Bonzo weigh the most?

____

3. What happened to Bonzo's weight between June and July?

____

4. Bonzo gained 5 pounds between March and April. How much did Bonzo weigh in March?

________ pounds

Use with Grade 2, Chapter 27, Lesson 4, pages 515–516.

Name ____

Problem Solving Skill: Reading for Math

Important and Unimportant Information

The gym teacher brought 5 different types of balls for children to use. She brought them in a blue bag. The graph below shows the type and number of balls she brought.

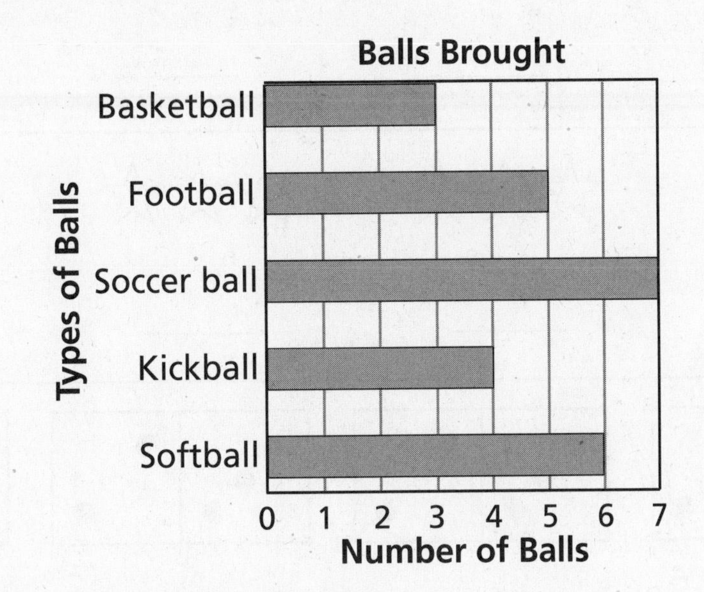

1. How many soccer balls are there? ______

2. How many basketballs are there? ______

3. Are there more footballs or kickballs? ____________

   How many more? ______

4. Which kind of ball are there fewest of? ____________

5. Which kind of ball are there the most of? ____________

6. What information in the story did not help you answer the questions? ____________________

Name ____________________

Explore Equal Groups

P 28-1 PRACTICE

Skip-count. Write how many in all.

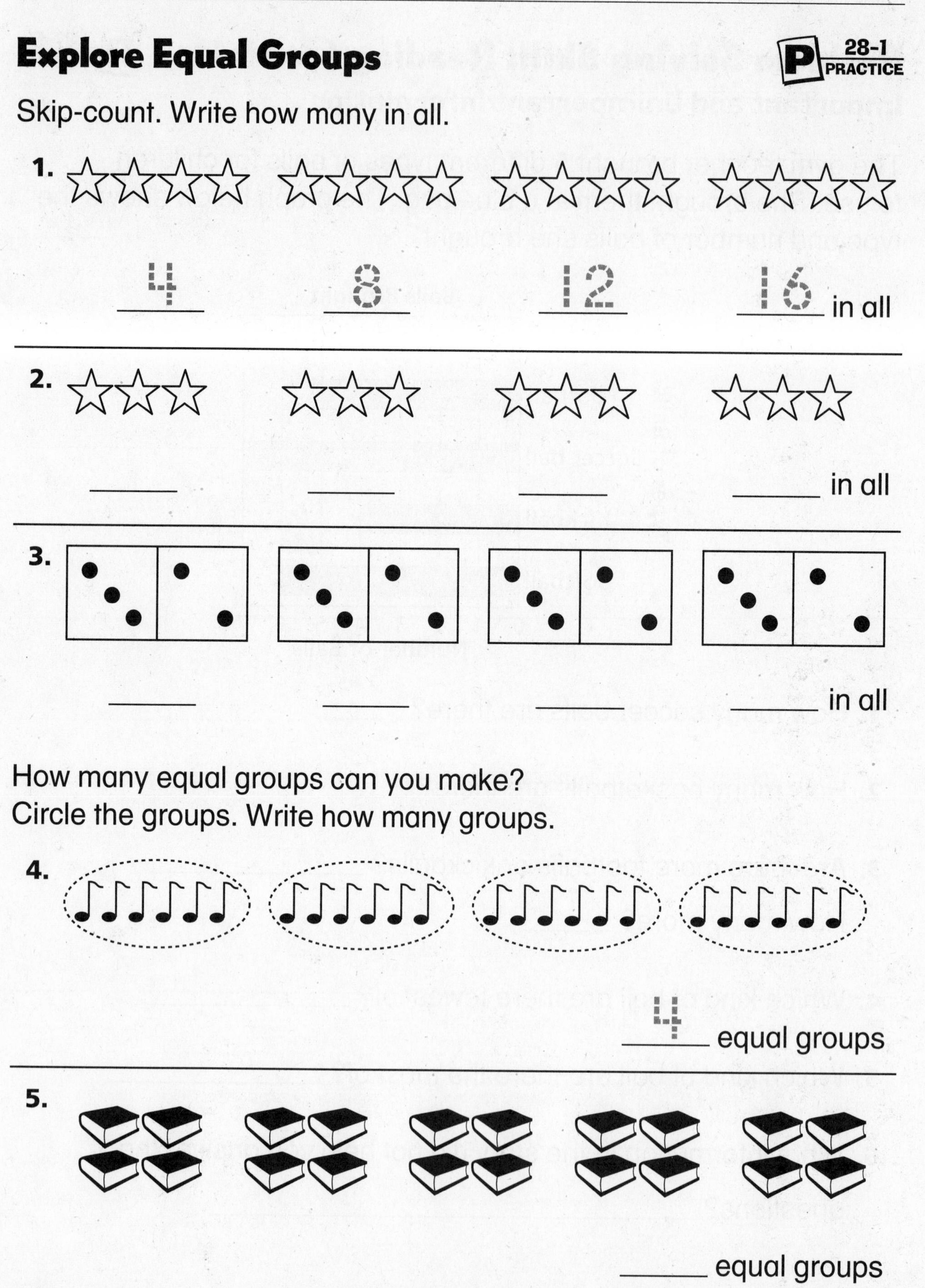

1. 4 8 12 16 in all

2. ______ ______ ______ ______ in all

3. ______ ______ ______ ______ in all

How many equal groups can you make?
Circle the groups. Write how many groups.

4. 4 equal groups

5. ______ equal groups

Use with Grade 2, Chapter 28, Lesson 1, pages 525–526.

Name ____________________

Repeated Addition and Multiplication

P 28-2 PRACTICE

Add. Then multiply.

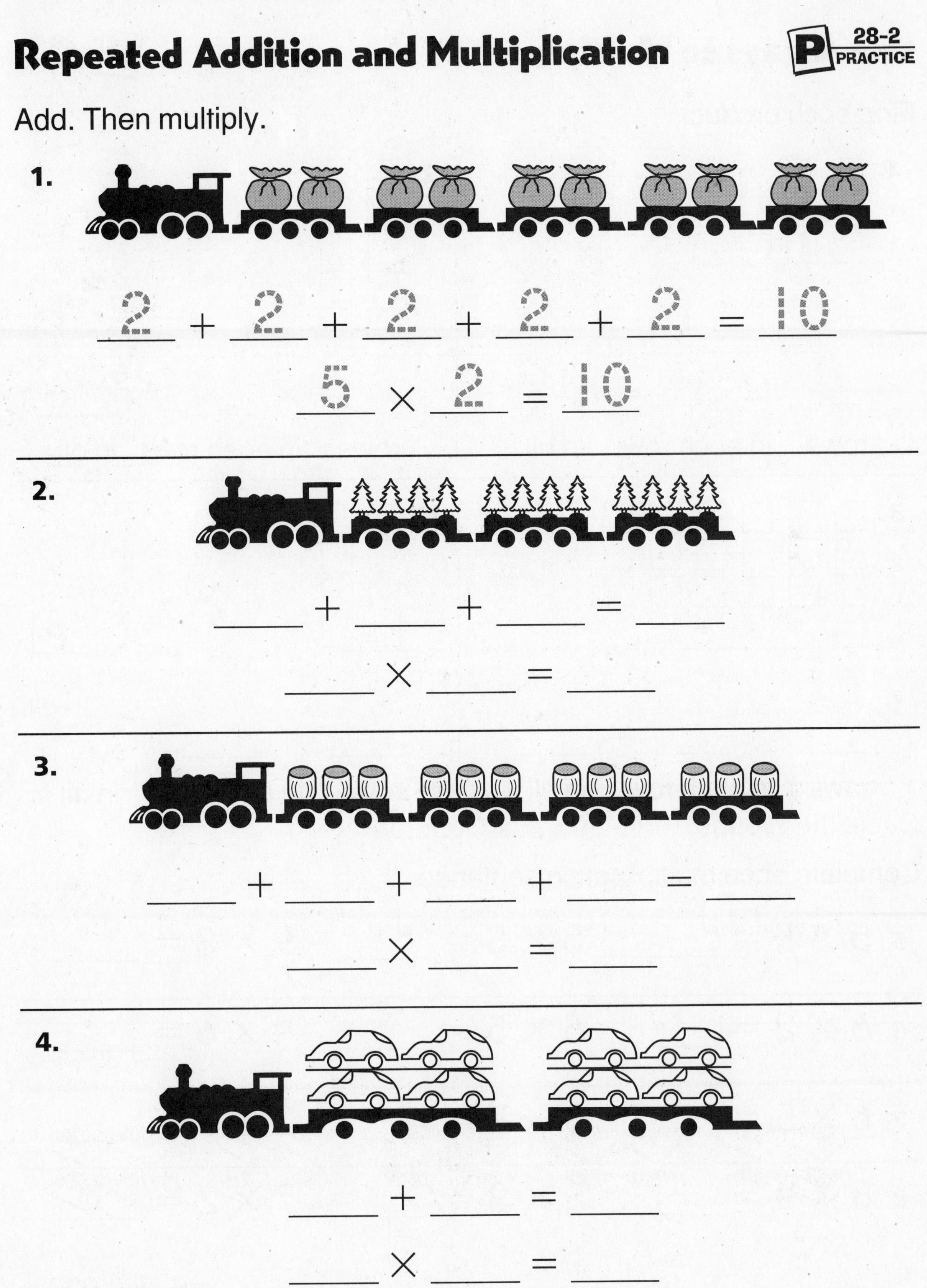

1. 2 + 2 + 2 + 2 + 2 = 10

5 × 2 = 10

2. ___ + ___ + ___ = ___

___ × ___ = ___

3. ___ + ___ + ___ + ___ = ___

___ × ___ = ___

4. ___ + ___ = ___

___ × ___ = ___

Name ______________________________

Use Arrays to Multiply

P 28-3 PRACTICE

Find each product.

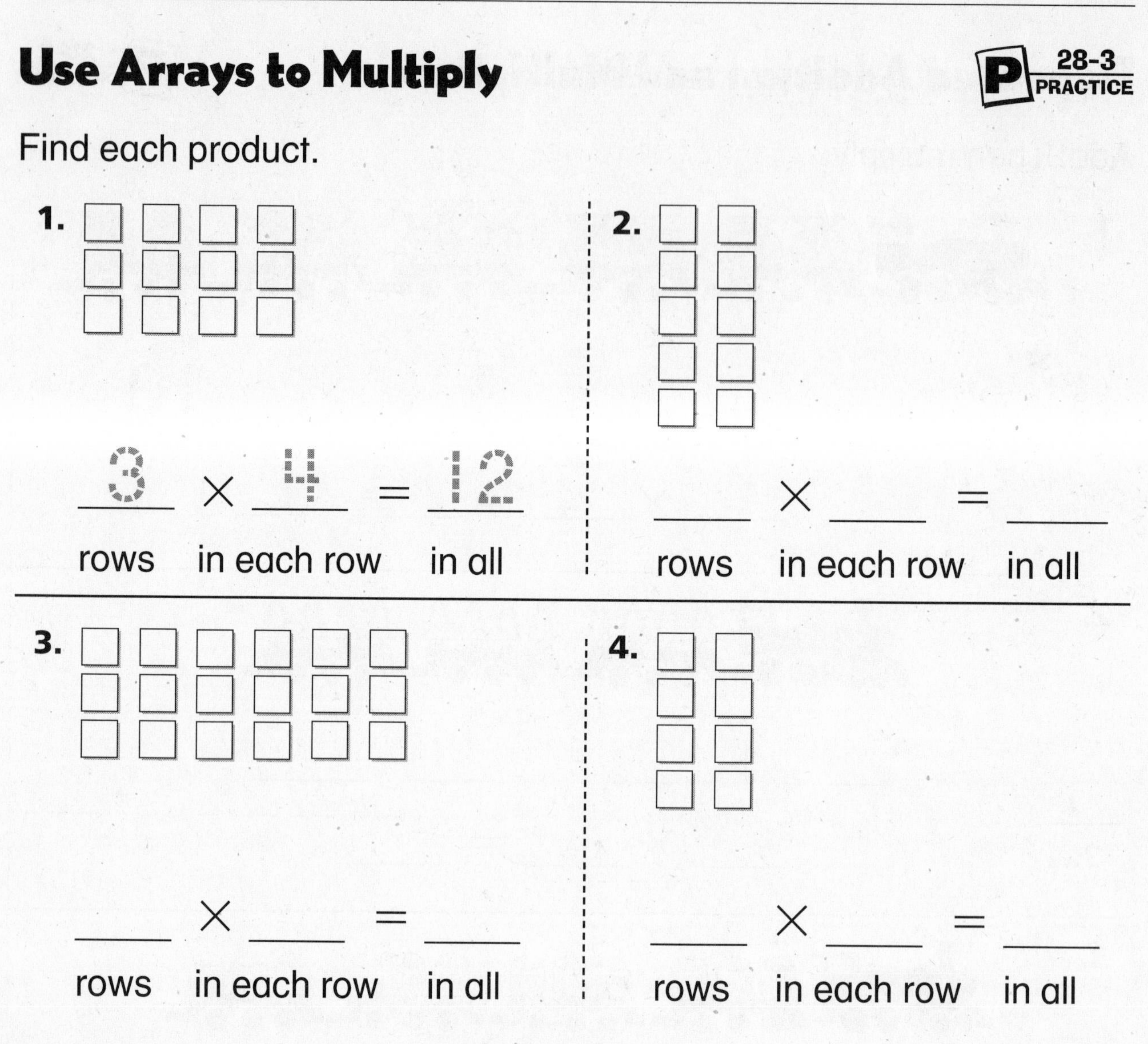

1. 3 × 4 = 12

rows in each row in all

2. ____ × ____ = ____

rows in each row in all

3. ____ × ____ = ____

rows in each row in all

4. ____ × ____ = ____

rows in each row in all

Complete each multiplication sentence.

5. $5 \times 4 =$ ____ $3 \times 8 =$ ____ $9 \times 3 =$ ____

6. $6 \times 2 =$ ____ $5 \times 5 =$ ____ $4 \times 6 =$ ____

7. $6 \times 5 =$ ____ $3 \times 7 =$ ____ $2 \times 5 =$ ____

8. $8 \times 4 =$ ____ $5 \times 3 =$ ____ $7 \times 2 =$ ____

Use with Grade 2, Chapter 28, Lesson 3, pages 529–530.

Name ______________________________

Repeated Subtraction and Division

Divide. You can draw a picture to help.

1. There are 20 people.
There are 5 on each team.
How many teams can you have?

$20 \div 5 =$ 4

4 teams

2. 15 people
3 in each car
How many cars?

$15 \div 3 =$ ______

______ cars

3. 28 days
7 days in each week
How many weeks?

$28 \div 7 =$ ______

______ weeks

Problem Solving

Draw a picture to solve.

4. Laura has 16 cookies to give to 4 friends. How many cookies does each friend get?

______ cookies

Show Your Work

Name ______________________________

Divide to Find Equal Shares

Divide. You can draw a picture to help.

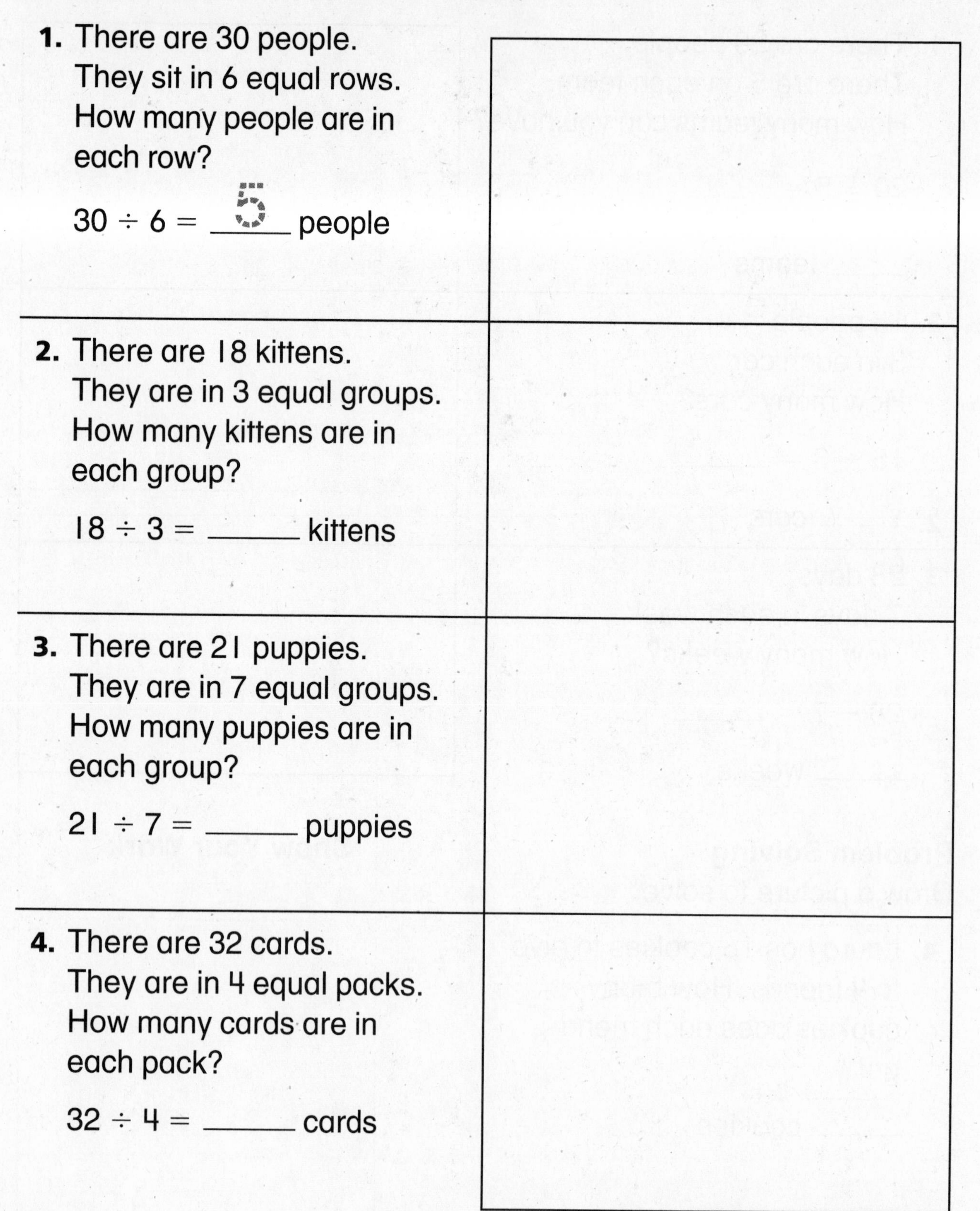

1. There are 30 people.
They sit in 6 equal rows.
How many people are in each row?

$30 \div 6 =$ 5 people

2. There are 18 kittens.
They are in 3 equal groups.
How many kittens are in each group?

$18 \div 3 =$ ______ kittens

3. There are 21 puppies.
They are in 7 equal groups.
How many puppies are in each group?

$21 \div 7 =$ ______ puppies

4. There are 32 cards.
They are in 4 equal packs.
How many cards are in each pack?

$32 \div 4 =$ ______ cards

Name ________________________________

Problem Solving: Strategy

Use a Pattern • Algebra

Use a number pattern to solve. Draw or write to explain.

1. How many cans are there in five 6-packs of juice?

Number of 6-packs	1	2	3	4	5
Number of cans	6	12	18	24	30

There are ______ cans in five 6-packs of juice. What pattern do I see?

____

2. There are 5 cats. How many legs are there in all?

Number of cats	1	2	3	4	5
Number of legs					

There are ______ legs in all.
What pattern do I see?

____

3. An octopus has 8 legs. There are 3 octopuses. How many legs are there in all?

______ legs

Summer Skills Refresher

Summer Skills

Green Turtles

At the aquarium, there are many green turtles.

1. A green turtle can lay 118 eggs. Show the number as hundreds, tens, and ones. Then, write the number in expanded form. Finally, write the word name.

   ______ hundreds ______ tens ______ ones

   ______ + ______ = ______

   ____

2. Green turtles live about 62 years. Show the number as tens and ones. Then, write it in expanded form. Finally, write the word name.

   ______ tens ______ ones

   ______ + ______ = ______

   ____

3. Last year, 153 green turtles were returned to a natural habitat. Show the number as hundreds, tens, and ones. Write it in expanded form and then write the word name.

   ______ hundreds ______ tens ______ ones

   ______ + ______ = ______

   ____

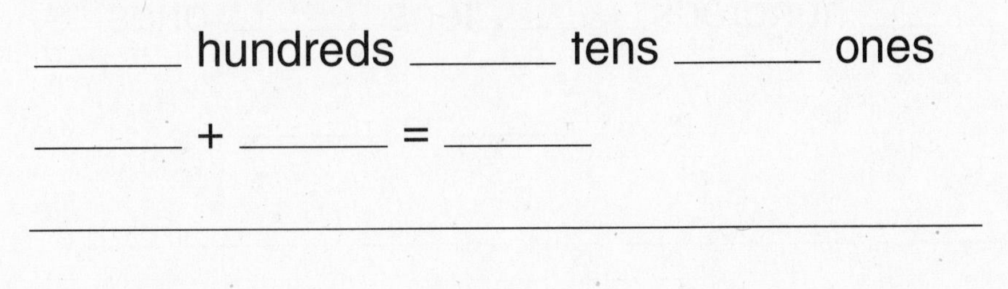

Answers: 1. 1 hundred 1 ten 8 ones; 100 + 10 + 8 = 118; one hundred eighteen; 2. 6 tens 2 ones; 60 + 2 = 62; sixty-two; 3. 1 hundred 5 tens 3 ones; 100 + 50 + 3 = 153; one hundred fifty-three

4. One turtle lays 107 eggs. A second turtle lays 75 eggs. How many eggs do the two turtles lay in all?

______ + ______ = ______

5. Show the number 174 as hundreds, tens and ones. Then, write it in expanded form and write the word name.

______ hundreds ______ tens ______ ones

______ + ______ + ______ = ______

____

6. The aquarium has two tanks. Each can hold 85 fish. How many fish can the two tanks hold?

______ + ______ = ______

7. Show the number 234 as hundreds, tens and ones. Then, write it in expanded form and write the word name.

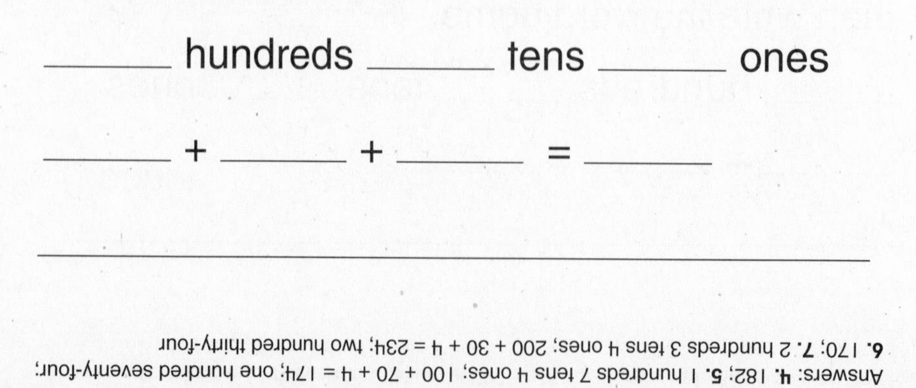

______ hundreds ______ tens ______ ones

______ + ______ + ______ = ______

____

Answers: 4. 182; 5. 1 hundreds 7 tens 4 ones; 100 + 70 + 4 = 174; one hundred seventy-four; 6. 170; 7. 2 hundreds 3 tens 4 ones; 200 + 30 + 4 = 234; two hundred thirty-four

Summer Skills

Measuring Turtles and Other Water Animals

Five of the seven species of sea turtles can be found along the coast of the United States.

1. How long is the green turtle in the drawing below? Estimate, then measure.

Estimate: _____ inches Measure: _____ inches

2. This is called the loggerhead turtle. How long is the turtle in the drawing below? Estimate, then measure.

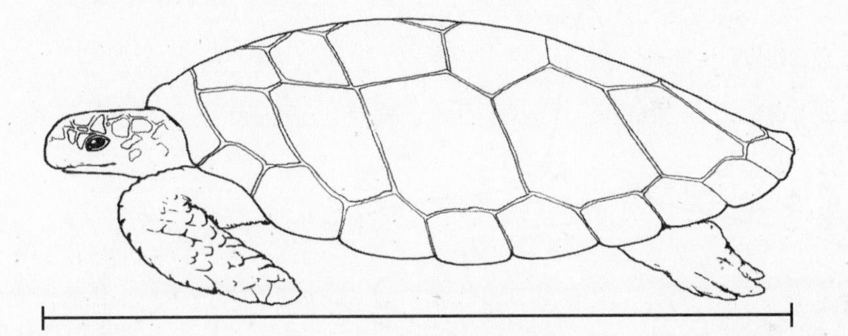

Estimate: _____ inches Measure: _____ inches

3. This is a drawing of the Atlantic ridley turtle. How long is the turtle in the drawing below? Estimate, then measure.

Estimate: _____ inches

Measure: _____ inches

Answers: 1. 3 inches; 2. 4 inches; 3. 2 inches

4. How long is the dolphin in the drawing below? Estimate, then measure.

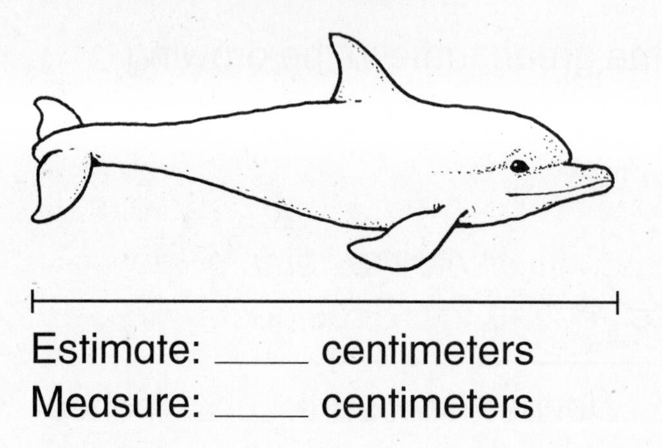

Estimate: _____ centimeters
Measure: _____ centimeters

5. How long is the killer whale in the drawing below? Estimate, then measure.

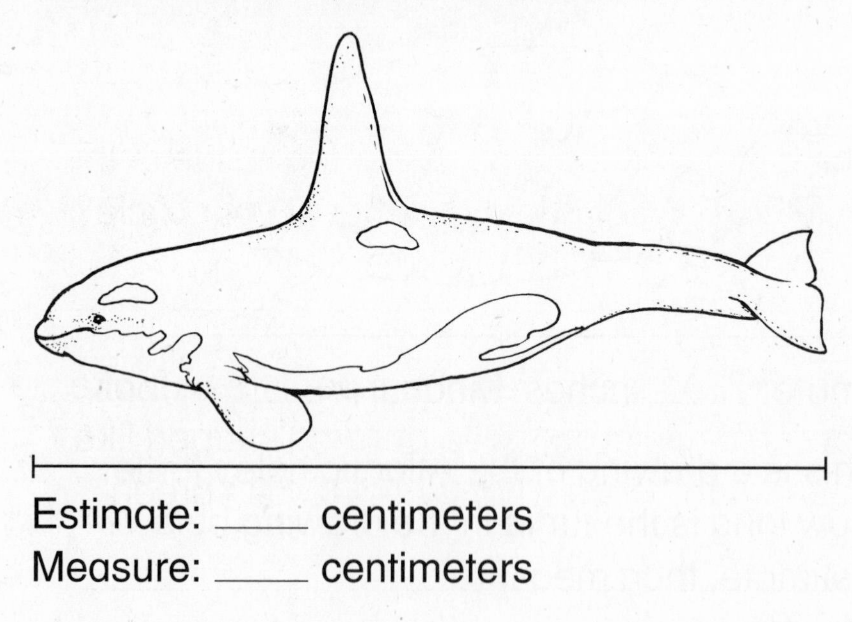

Estimate: _____ centimeters
Measure: _____ centimeters

Answers: **4.** Any estimate 7–9 centimeters; 8 centimeters; **5.** Any estimate 10–12 centimeters; 11 centimeters

Summer Skills

Out of This World

Florida is the home of the Kennedy Space Center at Cape Canaveral. The map shows the entrance.

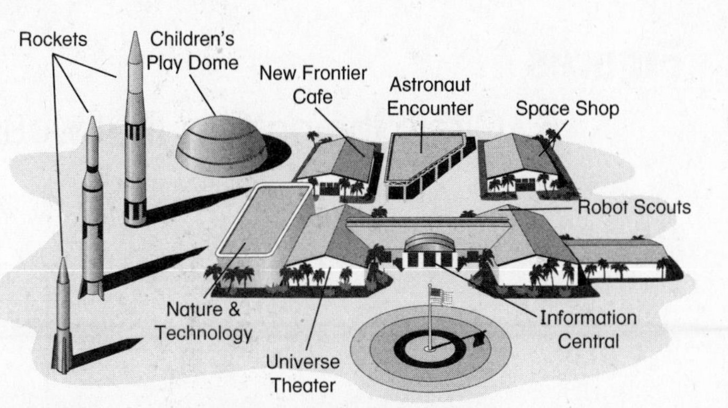

1. Find a rectangle on the map. Make a red circle around the rectangle. Write the name of the building where you found the rectangle.

   ____

2. Find a cylinder on the map. Make a yellow circle around the cylinder. Write what you circled in yellow.

   ____

3. Find a circle on the map. Color the inside of the circle blue. What is in the center of your circle?

   ____

4. One building is shaped like a pentagon. Make a green circle around the building shaped like a pentagon. Write what you circled in green.

   ____

Answers: 1. Accept any rooftop or the Nature & Technology building; 2. A rocket should have been circled or the flagpole; 3. The ground below the flagpole should be circled in blue; 4. The Astronaut Encounter should be circled. The ends of some buildings also.

Figures

5. Circle the pattern that would make a cylinder.

A. B.

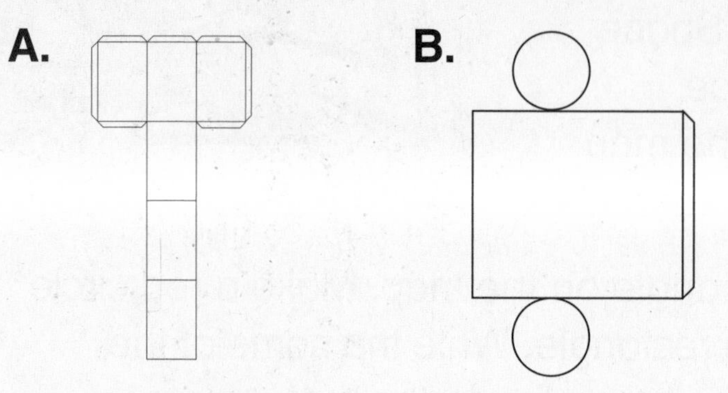

6. Circle the pattern that would make a rectangular prism.

A. B.

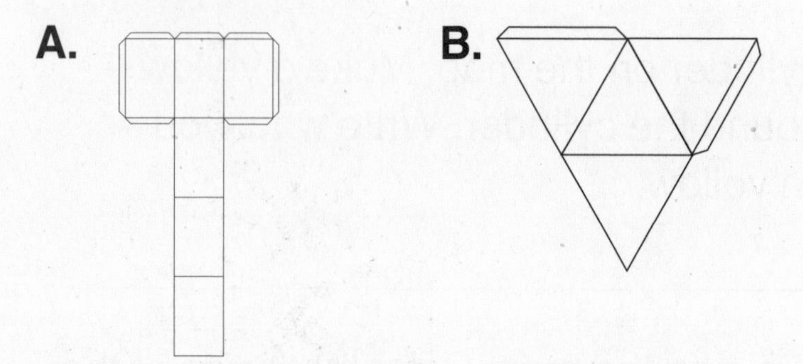

7. What figure would the pattern below make? Draw a picture of the figure and write the name.

____

Answers: 5. B; 6. A; 7. cube

Summer Skills

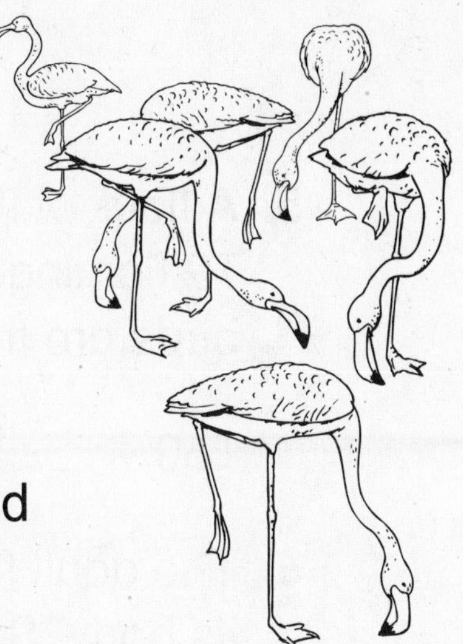

Flamingos

Flamingos are large pink birds. When flamingos rest, they stand on one leg.

1. Each of eight flamingos are standing on one leg. Eight of their legs are hidden from sight. Use doubles to find how many legs there are in all.

______ + ______ = ______

2. There are 30 flamingos in one flock. Skip-count by 2s to count to 30.

______, ______, ______, ______, ______,

______, ______, ______, ______, ______,

______, ______, ______, ______, ______

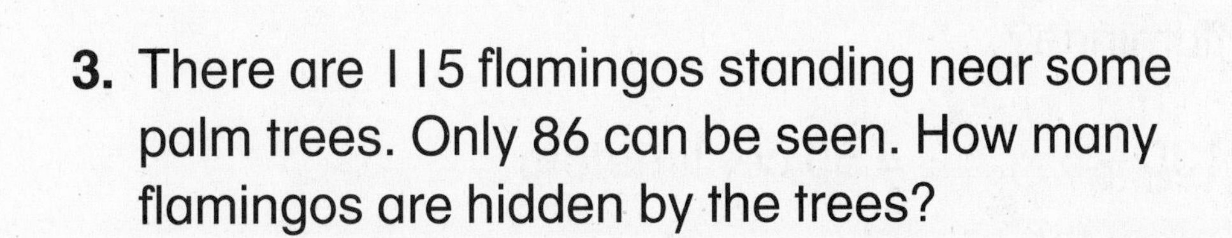

3. There are 115 flamingos standing near some palm trees. Only 86 can be seen. How many flamingos are hidden by the trees?

______ + 86 = 115

4. One flock of flamingos has 346 birds. Another flock has 237 birds in it. How many more birds are there in the larger flock? Find the missing addend.

______ + 237 = 346

Answers: **1.** 8 + 8 = 16; **2.** 2, 4, 6, 8, 10, 12, 14, 16, 18, 20, 22, 24, 26, 28, 30; **3.** 29; **4.** 109

5. A flock of flamingos has 160 birds. One day, 79 flamingos left the flock. How many fewer birds are there now? Find the difference.

160 – ______ = 79

6. The adult flamingo's legs are longer than its body. Some flamingos have legs that are 49 inches long. Other flamingos have legs that are 17 inches less than this. Find the length of the shorter legs.

______ + 17 = 49 inches

7. A tall flamingo can be 130 centimeters tall. A short flamingo can be 50 centimeters less than the taller one. What is the height of the shorter flamingo?

130 – ______ = 50 centimeters

8. The wingspan of some flamingos is 39 inches. Other flamingos can have a wingspan that is 65 inches. How much wider is the larger wingspan? Find the missing addend.

39 + ______ = 65 inches

Answers: **5.** 81; **6.** 32; **7.** 80; **8.** 26

Summer Skills

Beautiful Day

The weather in the southern part of the United States is pleasant most of the year. Use the graph to answer the questions.

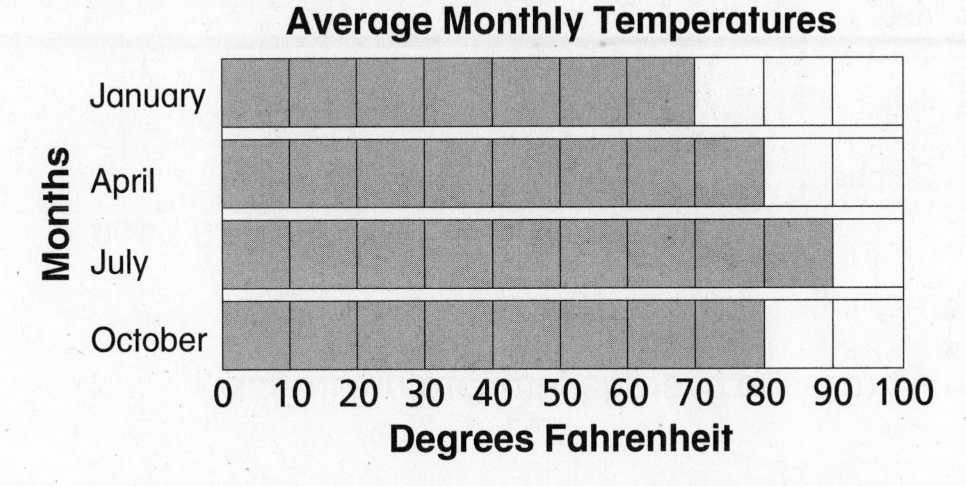

1. Which month has the coolest temperatures?

____

2. What is the average temperature during the coolest month?

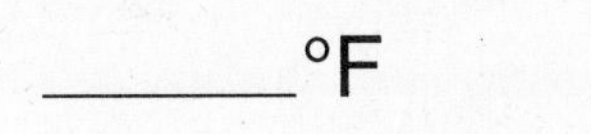

3. Which month is the warmest?

____

4. What is the average temperature during the warmest month?

______°F

Answers: 1. January; 2. 70°F; 3. July; 4. 90°F

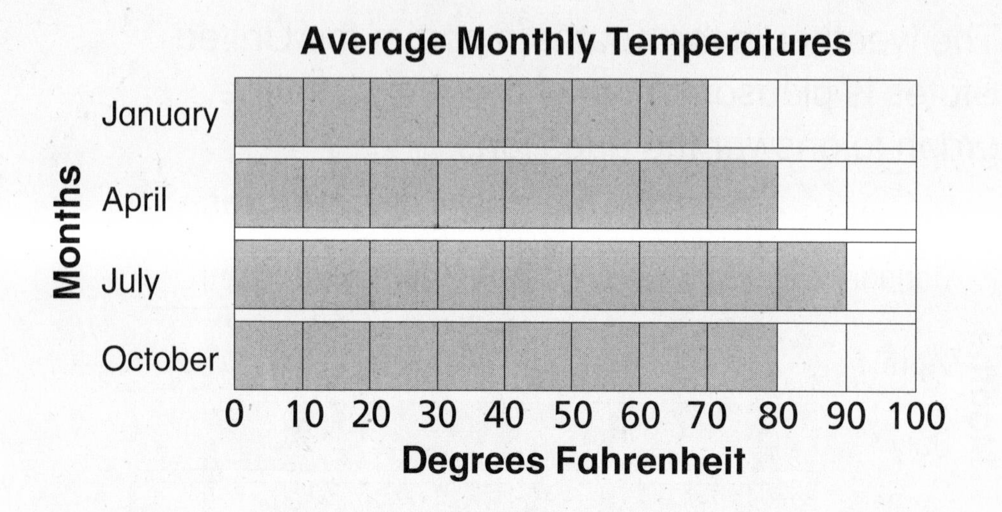

5. What is the average temperature in April?

______ °F

6. What is the average temperature in October?

______ °F

7. What is the range of the temperatures on the graph?

____

8. What is the mode of the temperatures on the graph?

____

Answers: **5.** 80°F; **6.** 80°F; **7.** 20°F; **8.** 80°.